10 Virginia SOL Grade 8 Math Practice Tests

The Ultimate Test Prep Collection with Answer Explanations

Dr. A. Nazari

10 Practice Tests

 *Grade 8 Mathematics* 

Welcome!

*This book contains **10 full-length practice tests** — the most comprehensive preparation you can get for your Grade 8 math assessment. Each test covers all six topics:*

Irrational Numbers   Powers & Scientific Notation

Linear Equations   Functions

Geometry   Data & Relationships

Ten tests give you the practice needed to walk into the real test feeling fully prepared.

Thorough preparation leads to outstanding results.

66 Ten full tests! By the time you finish, there won't be any surprises on test day. 99

 How to Use This Book

What's Inside

- **10 Full-Length Practice Tests** — each covers all 6 chapters of Grade 8 math: irrational numbers, exponents & scientific notation, linear equations, functions, geometry, and data analysis.
- **Detailed Answer Explanations** — every question includes a step-by-step solution so you learn from every mistake.
- **Formula Reference Sheet** — all the key Grade 8 formulas you need, organized and ready for quick review.
- **Test Tracker** — log your scores across all 10 tests and monitor your progress from start to finish.

Your 10-Test Training Plan

★ PHASE 1: Foundation (Tests 1–3)

Untimed or soft-timed. Focus on understanding the format, identifying strengths and weaknesses, and building good study habits.

★★ PHASE 2: Building Skills (Tests 4–7)

Timed (70 minutes each). Work on pacing, accuracy, and showing complete solutions. Review weak topics between tests.

★★★ PHASE 3: Test-Day Ready (Tests 8–10)

Full test conditions: strict timing, quiet space, no notes. Compare scores with your early tests to see your growth.

Schedule: Take one test every 3–4 days, or one per week. Use study days between tests to review.

Types of Questions

 **Multiple Choice:** *Four options — work the problem first, then match. Eliminate obviously wrong answers to narrow your choices.*

Short Answer & Constructed Response: *Show every step: equations, substitutions, simplifications. Partial credit rewards correct reasoning even if the final answer is off.*

Graphing & Data Analysis: *Plot points, draw lines, interpret graphs. Label axes clearly.*

Tip: Ten tests is a full preparation program. Don't rush. The key is what you do between tests — study, review, and understand your mistakes before moving forward.

Find more at
ViewMath.com/VA-Grade8

Test-Taking Tips

Your complete test-day toolkit

Before the Test

- Review your notes from the previous test — focus on your weak topics
- Set up a quiet, clean workspace with all your materials ready
- Start with a positive mindset: you've prepared for this

During the Test

- Read each problem fully before calculating anything
- Write the formula or set up the equation first, then substitute values
- Show all your work — every step, every operation
- If stuck for more than 2 minutes, mark it and move on
- Use estimation to check if your answers are reasonable

After the Test

- Read the full explanation for every question you got wrong
- Write down which topics gave you trouble (not just question numbers)
- Study those topics before taking the next test
- Record your score in the Test Tracker

Common Mistakes in Grade 8 Math

⚠ **Exponents:** $(ab)^n = a^n b^n$, but $a^m + a^n \neq a^{m+n}$. Only multiply/divide to combine.

⚠ **Slope formula:** $m = \frac{y_2 - y_1}{x_2 - x_1}$ — keep the order consistent.

⚠ **Systems of equations:** The solution must satisfy both equations.

⚠ **Transformations:** Rotations and reflections change position; dilations change size.

⚠ **Volume:** Use $\pi \approx 3.14$ or leave as π — match what the question asks.

The students who improve the most aren't the ones who take the most tests — they're the ones who carefully review every mistake. Make that your priority.

Find more at
ViewMath.com/VA-Grade8

What You'll Need

Gather these materials before you begin

📋 Materials Checklist

Sharpened Pencils — #2 pencils, at least two

Good Eraser — for clean corrections

Scratch Paper — for working out problems

Ruler / Straightedge — for graphing & geometry

Quiet Space — no distractions

Focused Mind — ready to do your best

✅ Allowed Materials

✔ Pencils and eraser

✔ Scratch paper (provided on official test day)

✔ Ruler or straightedge (if required)

✔ Protractor (if required)

🚫 Not Allowed

✖ Calculator (unless your state test allows it)

✖ Cell phone or any electronic device

✖ Notes, textbooks, or reference sheets

✖ Help from others during the test

Ten tests is a comprehensive program. Plan **one test every 3–4 days** (or one per week) with study sessions between each test.

How to help:

- Tests 1–3 should be untimed — build understanding before adding pressure.
- After each test, review the answer explanations together. Ask: "Which topics were hardest? Let's study those before the next one."
- Use the Test Tracker to celebrate progress over time.
- For topic-specific help, pair this book with our **Grade 8 Math Study Guide** or **Grade 8 Workbook**.

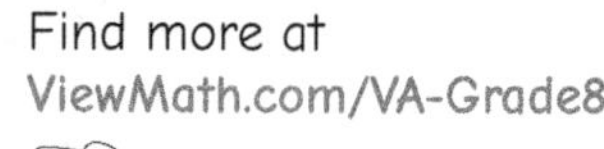

Find more at
ViewMath.com/VA-Grade8

Grade 8 Formula Reference

Keep this page handy — you may use it during your practice tests!

X^1 Exponent Rules

$$a^m \cdot a^n = a^{m+n} \qquad (a^m)^n = a^{mn} \qquad (ab)^n = a^n \cdot b^n$$

$$\frac{a^m}{a^n} = a^{m-n} \qquad a^0 = 1 \ (a \neq 0) \qquad a^{-n} = \frac{1}{a^n}$$

Lines & Linear Equations

Slope: $\ m = \dfrac{y_2 - y_1}{x_2 - x_1} = \dfrac{rise}{run}$
 m = slope   b = y-intercept

Slope-intercept: $\ y = mx + b$
 Parallel lines: same slope

Proportional: $\ y = mx$
 Proportional: passes through origin

Scientific Notation

$a \times 10^n \ $ where $1 \leq |a| < 10$ **Multiply:** *add exponents*   **Divide:** *subtract exponents*

$\sqrt{x}$ Roots & Number Sense

Perfect squares: 1, 4, 9, 16, 25, 36, 49, 64, 81, 100, 121, 144

Perfect cubes: 1, 8, 27, 64, 125 $\sqrt{2} \approx 1.414$ $\sqrt{3} \approx 1.732$ $\pi \approx 3.14159$

Pythagorean Theorem & Distance

$$a^2 + b^2 = c^2$$

c = hypotenuse (longest side of a right triangle)   **Distance:** $\ d = \sqrt{(x_2 - x_1)^2 + (y_2 - y_1)^2}$

Volume Formulas

Cylinder $\ V = \pi r^2 h$ **Cone** $\ V = \dfrac{1}{3}\pi r^2 h$ **Sphere** $\ V = \dfrac{4}{3}\pi r^3$

⬚ Angle Relationships

Triangle angle sum: 180°   **Exterior angle** = sum of two remote interior angles

Parallel lines + transversal: Alternate interior angles are equal • Co-interior angles sum to 180°

⬚ Functions

Each input → exactly one output   **Vertical line test:** if any vertical line hits graph more than once ⇒ not a function

Linear: constant rate of change $(y = mx + b)$ **Nonlinear:** rate of change varies

⟳ Transformations

Translation: slide   **Reflection:** flip   **Rotation:** turn   **Dilation:** resize

Congruent = same shape & size   Similar = same shape, proportional size

 **Tip:** Bookmark this page! Review it before each test so these formulas become second nature.

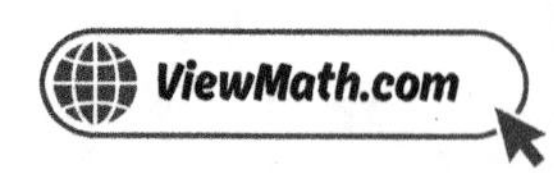

Multiplication Table

×	1	2	3	4	5	6	7	8	9	10	11	12
1	1	2	3	4	5	6	7	8	9	10	11	12
2	2	4	6	8	10	12	14	16	18	20	22	24
3	3	6	9	12	15	18	21	24	27	30	33	36
4	4	8	12	16	20	24	28	32	36	40	44	48
5	5	10	15	20	25	30	35	40	45	50	55	60
6	6	12	18	24	30	36	42	48	54	60	66	72
7	7	14	21	28	35	42	49	56	63	70	77	84
8	8	16	24	32	40	48	56	64	72	80	88	96
9	9	18	27	36	45	54	63	72	81	90	99	108
10	10	20	30	40	50	60	70	80	90	100	110	120
11	11	22	33	44	55	66	77	88	99	110	121	132
12	12	24	36	48	60	72	84	96	108	120	132	144

💡 How to Use This Table

To find **4 × 7**:

1. Find **4** in the left column (blue).
2. Find **7** in the top row (blue).
3. Follow the row and column until they meet: the answer is **28**!

> **Tip:** You can also use this table for division! If you know $28 \div 4 = ?$, find 28 in the 4's row. The column header gives you the answer: **7**!

Find more at
ViewMath.com/VA-Grade8

My Test Tracker

Record every test and watch your scores improve

Name: _______________________________   Start Date: _______________________

GETTING STARTED (Tests 1–3)

Test 1 — Untimed

Date: ______________   Score: ______ / ______   %: ______   Topics to review: _________________________

Test 2 — Untimed

Date: ______________   Score: ______ / ______   %: ______   Topics to review: _________________________

Test 3 — Soft Timer

Date: ______________   Score: ______ / ______   %: ______   Topics to review: _________________________

BUILDING SKILLS (Tests 4–7)

Test 4 — Timed (70 min)

Date: ______________   Score: ______ / ______   %: ______   Focus area: _________________________

Test 5 — Timed (70 min)

Date: ______________   Score: ______ / ______   %: ______   Focus area: _________________________

Test 6 — Timed (70 min)

Date: ______________   Score: ______ / ______   %: ______   Focus area: _________________________

Test 7 — Timed (70 min)

Date: ______________   Score: ______ / ______   %: ______   Focus area: _________________________

TEST-DAY READY (Tests 8–10)

Test 8 — Full Test Conditions

Date: ___________   Score: _______ / _______   %: _______   Growth since Test 1: _______________

Test 9 — Full Test Conditions

Date: ___________   Score: _______ / _______   %: _______   Growth since Test 1: _______________

Test 10 — Full Test Conditions

Date: ___________   Score: _______ / _______   %: _______   Growth since Test 1: _______________

📊 Score Progress

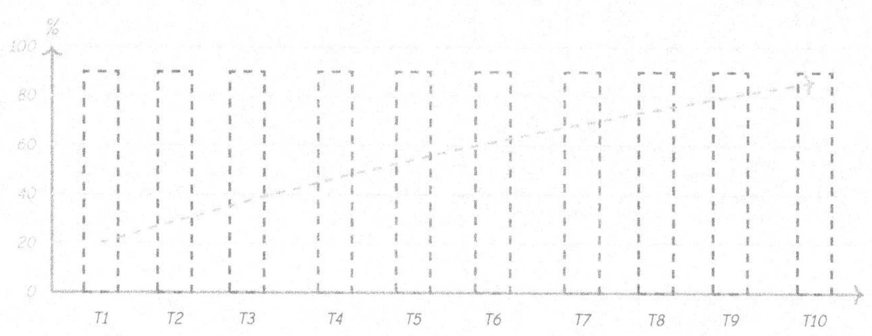

Shade each bar after every test. Watch your improvement!

✅ Final Reflection

The most important thing I learned: ____

The topic where I improved the most: _______________________________________

My advice for other students: ____

Find more at
ViewMath.com/VA-Grade8

⭐ Table of Contents ⭐

Here's what we'll explore together!

⭐ Practice Test 1 2

⭐ Practice Test 2 12

⭐ Practice Test 3 22

⭐ Practice Test 4 31

⭐ Practice Test 5 39

⭐ Practice Test 6 47

⭐ Practice Test 7 55

⭐ Practice Test 8 63

⭐ Practice Test 9 72

⭐ Practice Test 10 81

⭐ Answer Key & Explanations 93

Let's learn and have fun!

1

Practice Test 1

 30 Questions

---

✏️ Before You Start ✏️

✓ **Read each question carefully** before choosing your answer.

✓ **Show your work** on scratch paper when you need to.

✓ **Skip hard questions** and come back to them later.

✓ **Check your answers** when you're done.

✓ **Take your time** — there's no rush!

⭐ **You've Got This!** ⭐

Do your best and show what you know!

1. True or false: The number $\frac{22}{7}$ is equal to π.

   Your Answer:

2. The table below shows four repeating decimals and the power of 10 a student used to convert each. Which one uses the WRONG power of 10?

Decimal	Multiply by
$0.\overline{7}$	10
$0.\overline{45}$	100
$0.\overline{123}$	100
$0.\overline{8}$	10

(A) $0.\overline{7}$

(B) $0.\overline{45}$

(C) $0.\overline{123}$

(D) $0.\overline{8}$

3. The diagram below shows a rectangle with irrational side lengths. Which is the best estimate of the area?

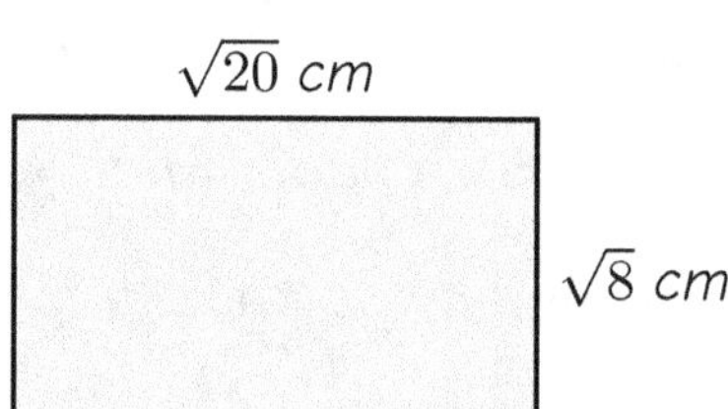

(A) $\sqrt{28} \approx 5.3$ cm^2

(B) $\sqrt{160} \approx 12.6$ cm^2

(C) 28 cm^2

(D) 160 cm^2

4. In the simple interest formula $I = Prt$, what does r represent?

(A) The total amount of money

(B) The interest rate as a decimal

(C) The time in months

(D) The principal investment

5. Study the number line below. Which expression corresponds to the point marked with a star ($\bigstar$)?

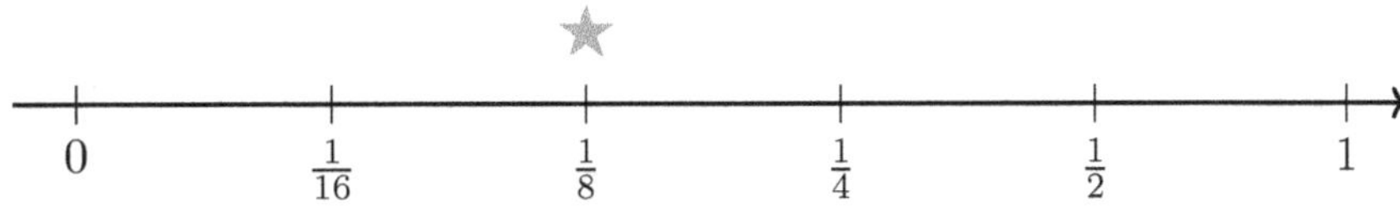

(A) 2^{-2}

(B) 2^{-3}

(C) 2^{-4}

(D) 2^{-1}

6. Estimate the product $498{,}000{,}000 \times 6{,}200$ by first writing each factor in scientific notation and then computing.

Your Answer:

7. Study the diagram below. Each box produces its output by multiplying the two inputs. What belongs in the output box marked "?"?

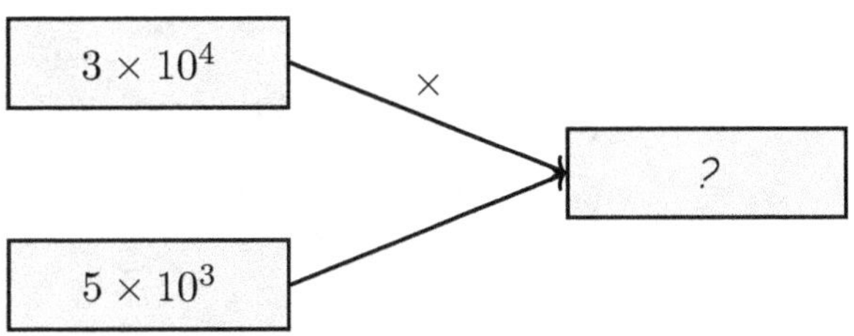

(A) 8×10^7

(B) 1.5×10^7

(C) 1.5×10^8

(D) 15×10^7

8. A bike rental company charges \$8 per hour. Which equation models the total cost y for x hours?

(A) $y = x + 8$

(B) $y = 8x$

(C) $y = \frac{x}{8}$

(D) $y = 8x + 10$

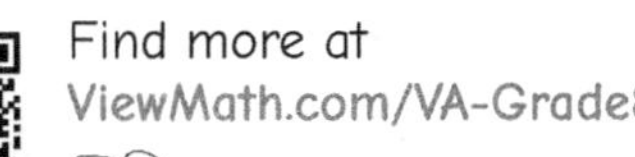

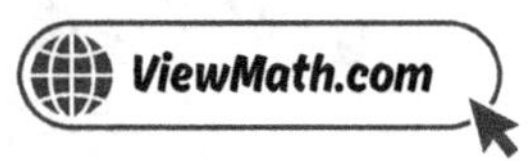

9. Solve $-5x + 14 = -x + 2$.

Your Answer:

10. A school play sold 200 tickets. Adult tickets were $8 and student tickets were $5. Total revenue was $1,240. How many adult tickets were sold?

Your Answer:

11. The graph shows two lines. Line A passes through $(0, 1)$ and $(3, 7)$. Line B passes through $(0, 4)$ and $(3, 7)$. Which statement is true?

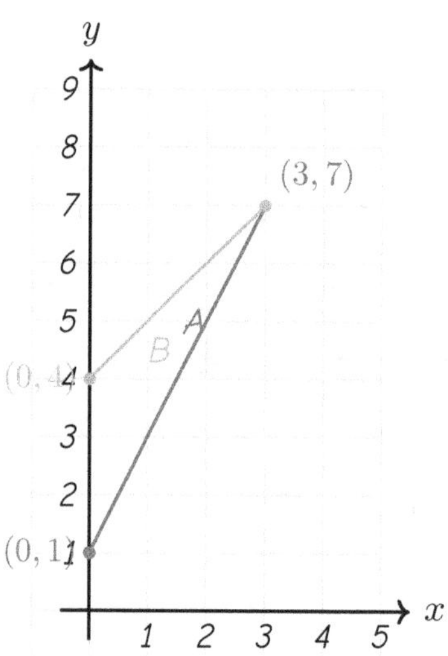

(A) Line A has a greater rate of change and a greater initial value.

(B) Line A has a greater rate of change but a smaller initial value.

(C) Line B has a greater rate of change and a greater initial value.

(D) Line B has a greater rate of change but a smaller initial value.

12. Look at the graph below. Is the function linear or nonlinear? Explain your reasoning.

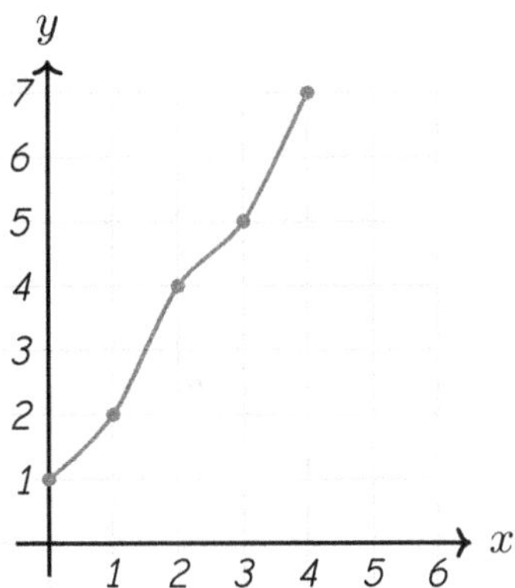

Your Answer:

13. A bathtub has 80 gallons of water and is draining at 5 gallons per minute. After how many minutes will it be empty?

Your Answer:

14. A graph of distance vs. time goes upward from left to right. What does this tell you?

(A) The object is slowing down.

(B) The distance is decreasing.

(C) The distance is increasing over time.

(D) The object is standing still.

15. Point $K(-3, -5)$ is rotated $90°$ counterclockwise around the origin. What are the coordinates of K'?

Your Answer:

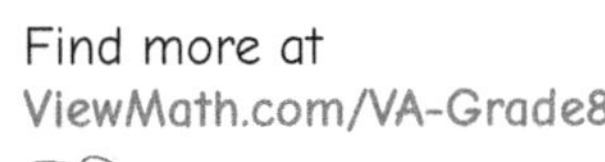

Find more at
ViewMath.com/VA-Grade8

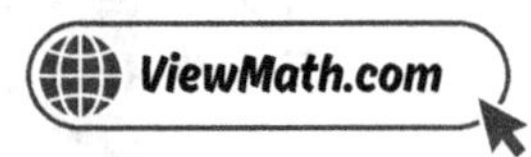

16. Which pair of figures is NOT necessarily congruent?

    (A) A triangle and its reflection

    (B) A rectangle and its translation

    (C) Two rectangles with the same perimeter

    (D) A square and its $90°$ rotation

17. Dilate the point $(-4, 10)$ by a factor of $\frac{1}{2}$ from the origin. What is the image?

    Your Answer:

18. A map has a scale factor of $\frac{1}{50,000}$. If a road is 3 cm on the map, how long is the real road in meters?

    Your Answer:

19. Parallel lines ℓ and m are cut by transversal t. If a pair of alternate interior angles are $(4x)°$ and $80°$, what is x?

    (A) 320

    (B) 20

    (C) 25

    (D) 10

20. Is the triangle with sides 9, 40, 41 a right triangle?

    (A) No, because $9 + 40 \neq 41$.

    (B) No, because $9^2 + 40^2 \neq 41^2$.

    (C) Yes, because $9^2 + 40^2 = 41^2$.

    (D) Yes, because $41 - 40 = 1$.

21. How many cones with the same base and height does it take to fill one cylinder?

    (A) 1

    (B) 2

    (C) 3

    (D) 4

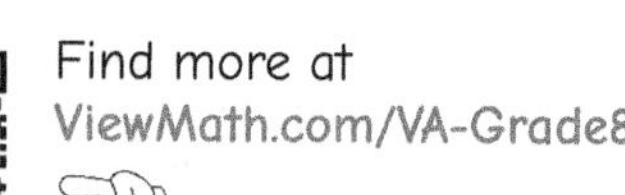

22. An angle and its complement are in the ratio 1 : 4. What is the larger angle?

(A) 18°

(B) 72°

(C) 36°

(D) 144°

23. What is the total surface area of a cylinder with radius 3 m and height 8 m? (Use $\pi \approx 3.14$.)

(A) 150.72 m^2

(B) 207.24 m^2

(C) 226.08 m^2

(D) 175.84 m^2

24. How many square pyramids with base side 4 cm and height 6 cm could fill a rectangular prism with dimensions $4 \times 4 \times 6$ cm?

(A) 1

(B) 2

(C) 3

(D) 4

25. A T-shaped figure is formed by a 12×2 cm rectangle on top and a 4×8 cm rectangle on the bottom. What is the total area?

(A) 48 cm^2

(B) 56 cm^2

(C) 64 cm^2

(D) 80 cm^2

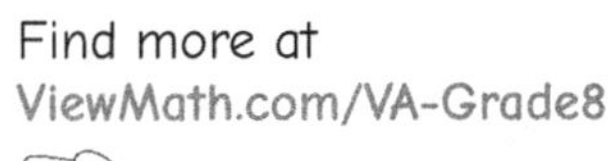
Find more at
ViewMath.com/VA-Grade8

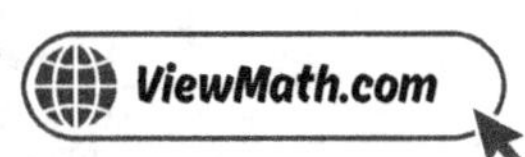

26. What type of association does this scatter plot show?

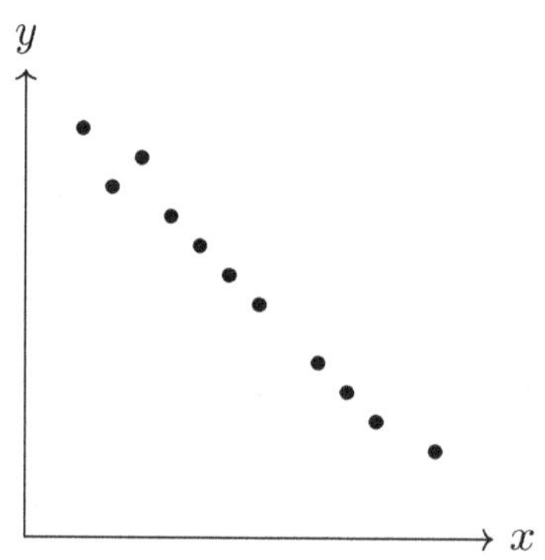

(A) Positive linear

(B) Negative linear

(C) No association

(D) Nonlinear

27. A scatter plot has 10 data points. One outlier is far above the trend. When drawing the line of best fit, you should:

(A) draw the line through the outlier

(B) ignore the outlier and fit the line to the other 9 points

(C) not draw a line at all

(D) draw the line only through the outlier

28. A trend line gives $y = -3x + 45$. Data was collected for $x = 1$ to $x = 10$. Predict y when $x = 5$ and when $x = 30$. State which prediction is more reliable and why.

Your Answer:

29. Using the table above, what percentage of 7th graders ride a bike?

(A) 55%

(B) 45%

(C) 62.5%

(D) 37.5%

30. *A deck has 52 cards. Two cards are drawn without replacement. Find* P(*both aces*).

Your Answer:

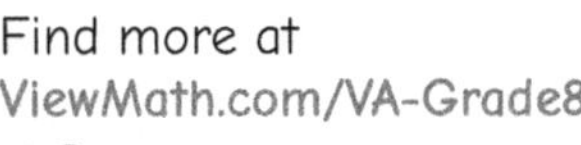

 # End of Practice Test 1 

Great job finishing the test!

☑ My Score

I got ___________ out of 30 questions right.

Check your answers in the **Answer Key** at the back of the book.

💡 Review any questions you missed. That's how we learn!

📊 Check Your Score Online!

Visit **ViewMath Academy** to enter your answers and see which topics you need to review. You can also explore lessons, take quizzes, track your scores, and save your progress!

viewmath.com/score/8.1.VA.16

Or go to *viewmath.com/score* and enter code: 8.1.VA.16

Practice Test 2

 30 Questions

✏️ Before You Start ✏️

- ✓ **Read each question carefully** before choosing your answer.
- ✓ **Show your work** on scratch paper when you need to.
- ✓ **Skip hard questions** and come back to them later.
- ✓ **Check your answers** when you're done.
- ✓ **Take your time** — there's no rush!

⭐ You've Got This! ⭐

Do your best and show what you know!

1. The number line below shows two points, P and Q.

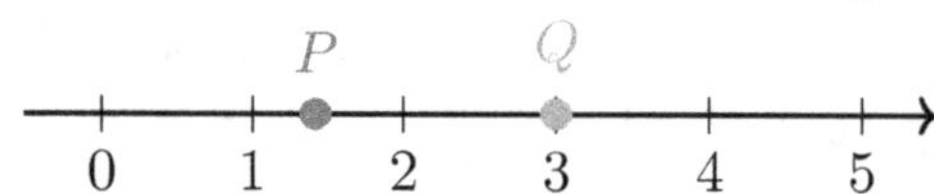

Point P represents $\sqrt{2}$ and point Q represents 3. Which statement is true?

(A) Both P and Q are rational.

(B) Both P and Q are irrational.

(C) P is irrational and Q is rational.

(D) P is rational and Q is irrational.

2. The flowchart below shows the conversion steps for a repeating decimal. Fill in the missing values to convert $0.\overline{54}$ to a fraction.

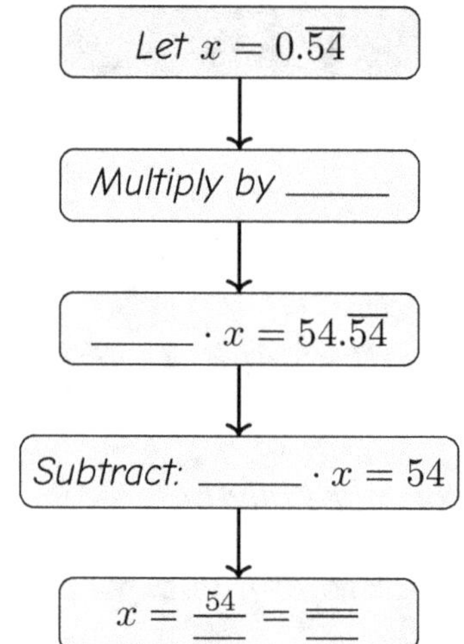

Your Answer

3. The number line below shows the values of π and $\sqrt{10}$ marked. Find the distance between them.

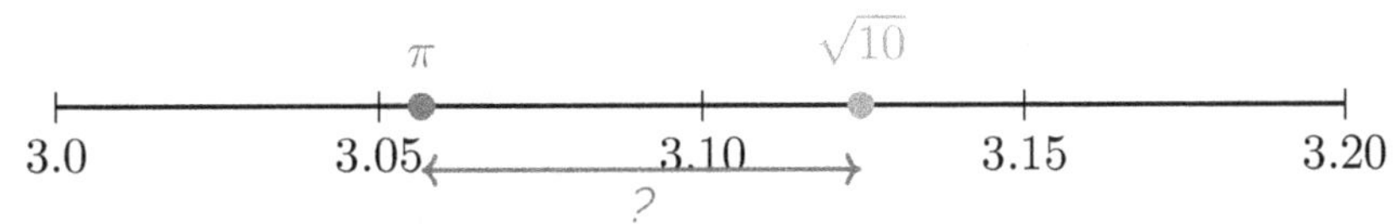

Your Answer:

4. A store marks up a \$50 item by 40%. What is the selling price?

(A) \$90

(B) \$70

(C) \$20

(D) \$54

5. Which expression is equivalent to $\frac{9^3 \cdot 9^2}{9^4}$?

(A) 9^{-1}

(B) 9^0

(C) 9^1

(D) 9^{24}

6. Which of the following is true about 4.2×10^{-3} and 4.2×10^3?

(A) They are equal

(B) 4.2×10^3 is 10^6 times as large as 4.2×10^{-3}

(C) 4.2×10^{-3} is 10^6 times as large as 4.2×10^3

(D) Their product is 1

7. What is $\frac{8 \times 10^7}{2 \times 10^3}$?

(A) 4×10^4

(B) 6×10^4

(C) 4×10^{10}

(D) 16×10^{10}

Find more at
ViewMath.com/VA-Grade8

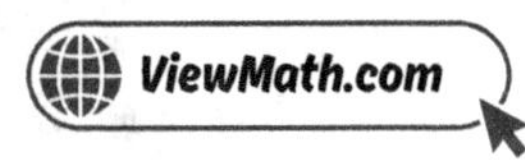

8. A faucet leaks at a constant rate. It leaks 45 mL in 15 minutes. How many mL does it leak in 1 hour?

Your Answer:

9. Solve $\frac{x}{4} + 3 = 7$.

(A) $x = 1$

(B) $x = 10$

(C) $x = 16$

(D) $x = 28$

10. You have \$5 bills and \$10 bills totaling \$85. You have 12 bills. How many \$10 bills do you have?

Your Answer:

11. Function A has a slope of $\frac{1}{2}$ and passes through the origin. Function B has a slope of $\frac{3}{4}$ and passes through the origin. Which is steeper?

(A) Function A

(B) Function B

(C) They have the same steepness.

(D) Cannot be determined.

12. A table shows $(1, 1)$, $(2, 8)$, $(3, 27)$, $(4, 64)$. Is this linear or nonlinear? What pattern do you see?

Your Answer:

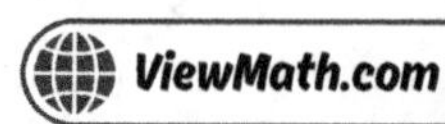

13. The table shows a real-world function. Write the equation and explain what m and b represent.

Hours worked (x)	Pay (y)
0	$50
1	$62
2	$74
3	$86

Your Answer:

14. A graph shows a line going upward with a constant slope. Is the rate of change constant or varying?

Your Answer:

15. Point $A(3, 5)$ is reflected over the x-axis. What are the coordinates of A'?

(A) $(-3, 5)$

(B) $(3, -5)$

(C) $(-3, -5)$

(D) $(5, 3)$

16. Triangle P has angles 40°, 60°, and 80°. Triangle Q has angles 40°, 60°, and 80°. Are they congruent?

(A) Yes, because matching angles guarantee congruence.

(B) No, because angles alone do not guarantee congruence.

(C) Yes, because all three angles match.

(D) No, because the angles are in different order.

17. Point $(3, 4)$ is dilated by a factor of 2 from the origin. What is the image?

(A) $(5, 6)$

(B) $(6, 8)$

(C) $(1.5, 2)$

(D) $(6, 4)$

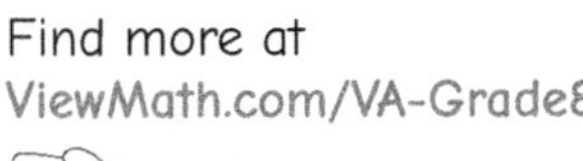

Find more at
ViewMath.com/VA-Grade8

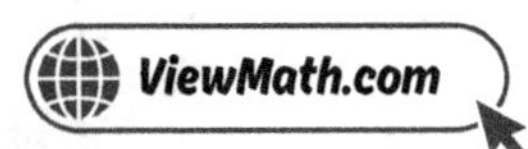

18. Two similar triangles have a scale factor of $\frac{1}{3}$ from the larger to the smaller. If the larger triangle's perimeter is 36 cm, what is the smaller triangle's perimeter?

(A) 12 cm

(B) 108 cm

(C) 9 cm

(D) 18 cm

19. An exterior angle of a triangle measures 118°. One of the non-adjacent interior angles is 43°. Find the other non-adjacent interior angle.

Your Answer

20. Is a triangle with sides 7, 24, 25 a right triangle?

(A) Yes, because $7 + 24 > 25$.

(B) Yes, because $7^2 + 24^2 = 25^2$.

(C) No, because $7^2 + 24^2 \neq 25^2$.

(D) No, because all sides must be equal.

21. A traffic cone has radius 8 in and height 24 in. What is its volume? Use $\pi \approx 3.14$.

(A) 4,823.0 in^3

(B) 1,607.7 in^3

(C) 803.8 in^3

(D) 602.9 in^3

22. Two adjacent supplementary angles measure $(5x + 25)°$ and $(3x + 11)°$. Find the larger angle.

Your Answer

23. A square pyramid has a base side of 10 ft and a slant height of 13 ft. What is its surface area?

(A) 360 ft^2

(B) 260 ft^2

(C) 100 ft^2

(D) 280 ft^2

Find more at
ViewMath.com/VA-Grade8

24. A triangular pyramid has a base with legs 5 cm and 12 cm (a right triangle) and a height of 10 cm. Find the volume.

Your Answer:

25. What is the area of the figure below?

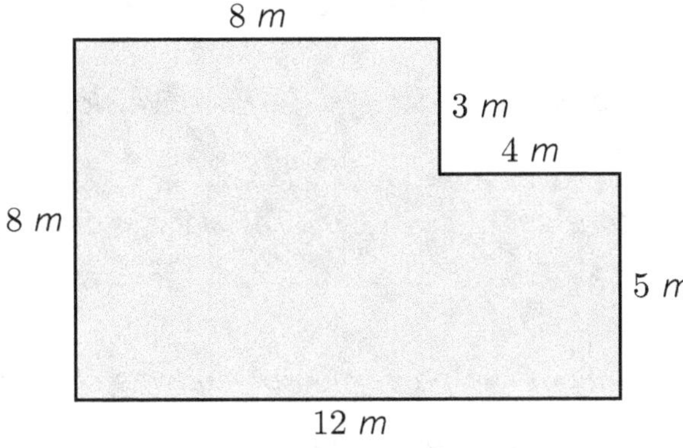

(A) 60 m²

(B) 84 m²

(C) 96 m²

(D) 76 m²

26. Describe the association shown in this scatter plot. Mention direction, shape, and any unusual features.

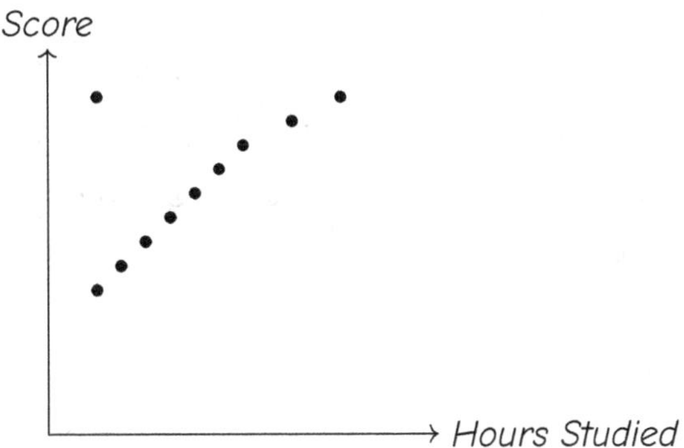

Your Answer:

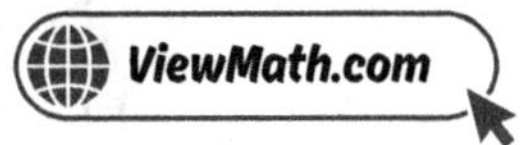

27. *What is a line of best fit?*

    (A) A line that passes through every data point

    (B) A line that connects the first and last data points

    (C) A straight line that comes close to most of the data points

    (D) A vertical line through the middle of the scatter plot

28. *Which statement about slope in a linear model is true?*

    (A) A positive slope means y decreases as x increases.

    (B) A negative slope means y increases as x increases.

    (C) A slope of zero means y stays constant.

    (D) The slope is always positive in real data.

29. *A segmented (stacked) bar chart shows the proportion of students choosing band or chorus by gender. The boys' bar is 60% band and 40% chorus. The girls' bar is 45% band and 55% chorus. What can you conclude?*

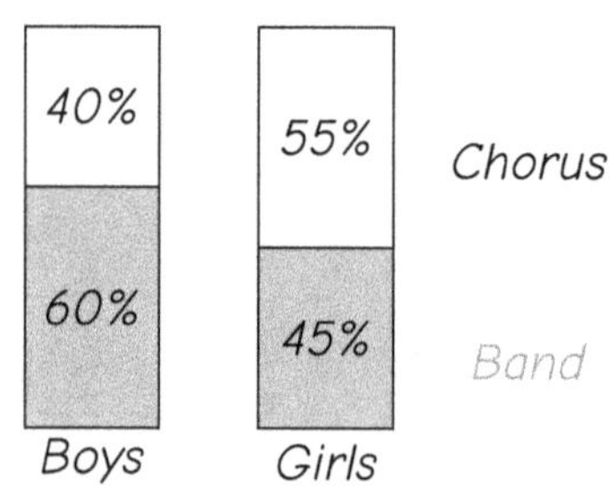

    (A) There is no association between gender and music choice.

    (B) Boys are more likely to choose band; girls more likely to choose chorus.

    (C) Girls and boys prefer band equally.

    (D) Chorus is more popular overall.

30. A tree diagram shows flipping a coin and rolling a die (1–6). How many total outcomes are in the sample space?

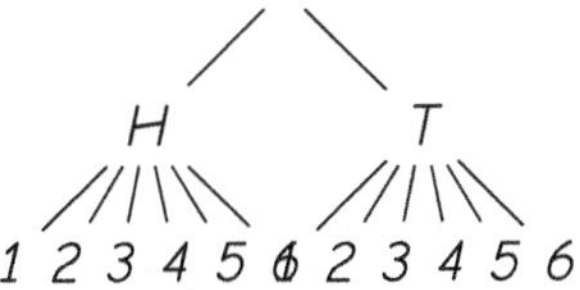

(A) 6

(B) 8

(C) 12

(D) 36

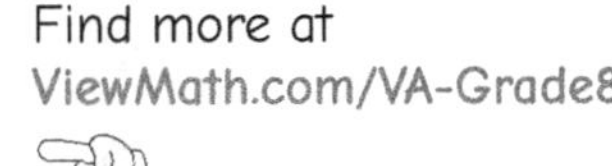

Find more at
ViewMath.com/VA-Grade8

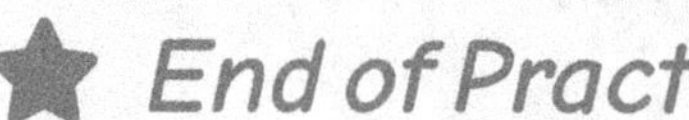 End of Practice Test 2 

Great job finishing the test!

 My Score

I got _____________ out of 30 questions right.

*Check your answers in the **Answer Key** at the back of the book.*

💡 *Review any questions you missed. That's how we learn!*

📊 Check Your Score Online!

Visit **ViewMath Academy** to enter your answers and see which topics you need to review. You can also explore lessons, take quizzes, track your scores, and save your progress!

viewmath.com/score/8.1.VA.17

Or go to viewmath.com/score and enter code: 8.1.VA.17

Practice Test 3

 *30 Questions*

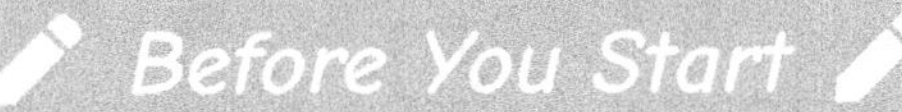

✏️ Before You Start ✏️

- ✔ **Read each question carefully** before choosing your answer.
- ✔ **Show your work** on scratch paper when you need to.
- ✔ **Skip hard questions** and come back to them later.
- ✔ **Check your answers** when you're done.
- ✔ **Take your time** — there's no rush!

 ⭐ You've Got This! ⭐ 

Do your best and show what you know!

1. Which number below can be written as a fraction $\frac{a}{b}$ where a and b are integers and $b \neq 0$?

  (A) $\sqrt{3}$ (B) π

  (C) $0.\overline{81}$ (D) $\sqrt{11}$

2. What is $0.\overline{27}$ as a fraction in simplest form?

  (A) $\frac{27}{100}$ (B) $\frac{27}{99}$

  (C) $\frac{3}{11}$ (D) $\frac{9}{33}$

3. Which is larger: $4\sqrt{2}$ or $\sqrt{30}$?

  (A) $4\sqrt{2}$, because $4 > \sqrt{30}$ (B) $\sqrt{30}$, because $30 > 2$

  (C) $4\sqrt{2}$, because $4\sqrt{2} \approx 5.66 > 5.48 \approx \sqrt{30}$ (D) They are equal.

4. Jake borrows \$2,000 at 6% simple interest for 2 years. How much total does he pay back?

  (A) \$2,120   (B) \$2,240

  (C) \$2,600   (D) \$2,012

5. A student simplifies $2^3 \cdot 2^4$ and gets 4^7. What mistake did the student make?

  (A) Added the exponents instead of multiplying them   (B) Multiplied the bases instead of keeping the same base

  (C) Subtracted the exponents instead of adding them   (D) Forgot to apply the zero exponent rule

Find more at
ViewMath.com/VA-Grade8

6. Which of the following is 0.00072 written in scientific notation?

(A) 7.2×10^{-4}

(B) 7.2×10^4

(C) 72×10^{-5}

(D) 0.72×10^{-3}

7. A virus has a mass of 9.5×10^{-18} grams. If a sample contains 2×10^6 viruses, what is the total mass?

(A) 1.9×10^{-11} g

(B) 1.9×10^{-12} g

(C) 11.5×10^{-12} g

(D) 1.9×10^{-24} g

8. Store A sells apples for $2 per pound. Store B's prices are shown in the table.

Pounds	3	6	9
Cost ($)	7.50	15.00	22.50

Which store has the lower unit price?

(A) Store A

(B) Store B

(C) They charge the same price.

(D) Not enough information.

9. Solve $0.5x + 1.5 = 4$

(A) $x = 3$

(B) $x = 5$

(C) $x = 7$

(D) $x = 11$

10. Two trains leave the same station heading in opposite directions. Train A goes 60 mph and Train B goes 80 mph. After how many hours are they 420 miles apart?

(A) 2

(B) 2.5

(C) 3

(D) 3.5

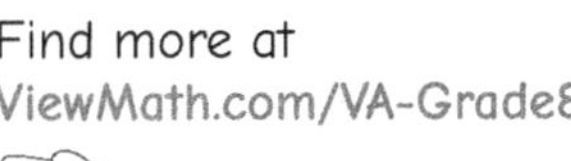

Find more at
ViewMath.com/VA-Grade8

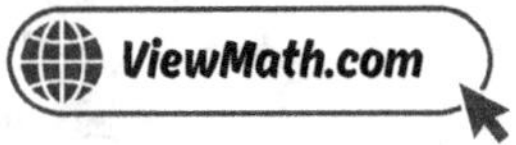

11. Function R from the previous table has a rate of change of:

x	0	2	4	6
y	1	9	17	25

(A) 2

(B) 3

(C) 4

(D) 8

12. A function increases by 6 for every increase of 1 in x. Is it linear or nonlinear?

Your Answer:

13. The function $y = 50x + 200$ models a bank account balance after x weeks. What is the weekly deposit?

(A) $50

(B) $100

(C) $200

(D) $250

14. Describe the graph of a cup of hot coffee cooling down to room temperature over an hour.

Your Answer:

15. Which transformation turns a figure around a fixed point?

(A) Translation

(B) Reflection

(C) Rotation

(D) Dilation

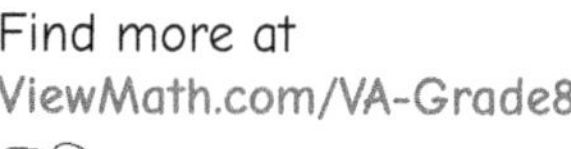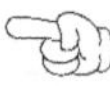

Find more at
ViewMath.com/VA-Grade8

ViewMath.com

16. Two figures have the same shape but one is twice as large. Are they congruent?

(A) Yes, same shape means congruent.

(B) Yes, because their angles match.

(C) No, they are similar but not congruent.

(D) No, they are neither similar nor congruent.

17. Point $P(-3, 4)$ is shown on the grid. It is transformed to $P'(6, 8)$. Which transformation was applied?

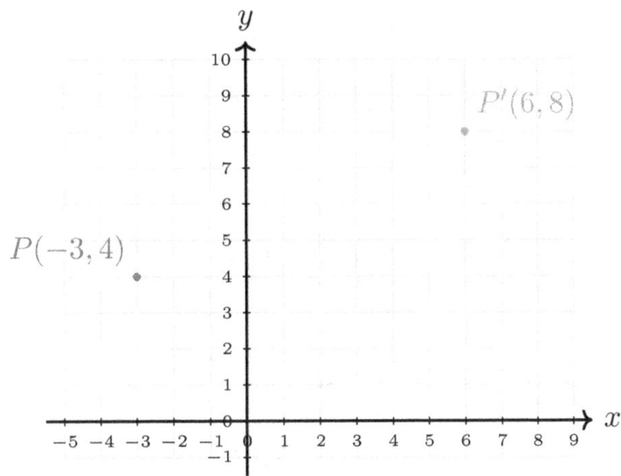

(A) Translation $(9, 4)$

(B) Dilation by factor 2

(C) Reflection over the y-axis

(D) Dilation by factor -2

18. A dilation with $k = 1$ produces:

(A) A figure twice as large

(B) A congruent figure

(C) A figure half as large

(D) No figure at all

19. Two angles of a triangle are $37°$ and $53°$. What is the exterior angle adjacent to the third angle?

(A) $90°$

(B) $127°$

(C) $53°$

(D) $143°$

20. A right triangle has hypotenuse 17 and one leg 8. What is the other leg?

(A) 9

(B) 15

(C) 25

(D) $\sqrt{225}$

21. A cylinder has volume 200π cm^3 and height 8 cm. What is the radius?

(A) 25 cm

(B) 5 cm

(C) $\sqrt{25}$ cm

(D) 10 cm

22. Find the value of x in the figure below where two lines intersect.

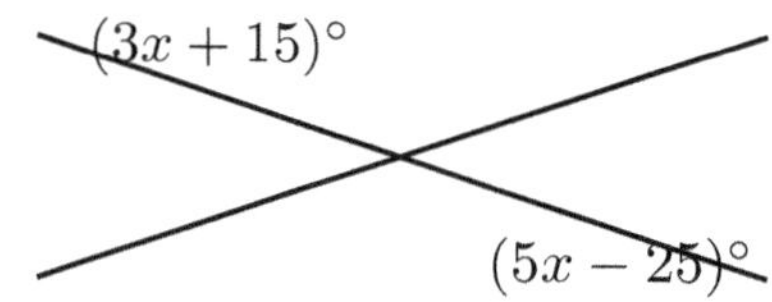

Your Answer

23. A square pyramid has total surface area 360 in^2 and a base side of 12 in. What is the slant height?

(A) 7 in

(B) 9 in

(C) 10 in

(D) 14 in

24. A square pyramid has base side 7 cm and height 12 cm. What is the volume?

(A) 196 cm^3

(B) 588 cm^3

(C) 294 cm^3

(D) 98 cm^3

Find more at
ViewMath.com/VA-Grade8

ViewMath.com

25. What is the best first step when finding the area of a composite figure?

(A) Multiply all given dimensions.

(B) Decompose it into simpler shapes whose areas you can compute.

(C) Find the perimeter first.

(D) Use the Pythagorean theorem.

26. Can a scatter plot show a positive association that is nonlinear? Give an example.

Your Answer:

27. A scatter plot shows a strong nonlinear (curved) pattern. Should you draw a straight line of best fit?

(A) Yes — always draw a straight line.

(B) No — a straight line does not fit curved data well.

(C) Yes — but only if there are outliers.

(D) No — you should never draw any line.

28. Using the model $y = -4x + 100$, predict y when $x = 10$.

(A) 140

(B) 60

(C) 96

(D) 40

29.

	Dog	Cat	Total
Boys	15	10	25
Girls	12	13	25
Total	27	23	50

How many girls prefer dogs?

(A) 10

(B) 12

(C) 13

(D) 15

Find more at
ViewMath.com/VA-Grade8

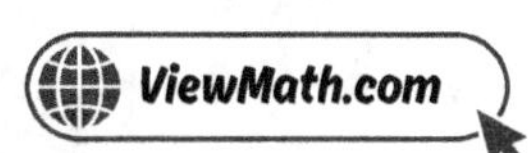

30. A fair die is rolled twice. What is $P(\text{both rolls are } 6)$?

(A) $\frac{1}{6}$

(B) $\frac{1}{12}$

(C) $\frac{1}{36}$

(D) $\frac{2}{6}$

Find more at
ViewMath.com/VA-Grade8

⭐ End of Practice Test 3 ⭐

Great job finishing the test!

 My Score

I got _____________ out of 30 questions right.

Check your answers in the **Answer Key** at the back of the book.

💡 Review any questions you missed. That's how we learn!

 Check Your Score Online!

Visit **ViewMath Academy** to enter your answers and see which topics you need to review. You can also explore lessons, take quizzes, track your scores, and save your progress!

viewmath.com/score/8.1.VA.18

Or go to viewmath.com/score and enter code: 8.1.VA.18

Practice Test 4

 30 Questions

✏ Before You Start ✏

- ✔ **Read each question carefully** before choosing your answer.
- ✔ **Show your work** on scratch paper when you need to.
- ✔ **Skip hard questions** and come back to them later.
- ✔ **Check your answers** when you're done.
- ✔ **Take your time** — there's no rush!

 ⭐ You've Got This! ⭐ 

Do your best and show what you know!

1. *Which of the following numbers is irrational?*

   (A) $\frac{5}{8}$

   (B) $0.\overline{6}$

   (C) $\sqrt{9}$

   (D) $\sqrt{7}$

2. *Write $0.5\overline{8}$ as a fraction in simplest form.*

   Your Answer:

3. *A student claims that $2\sqrt{5} = \sqrt{10}$. Is this correct?*

   (A) Yes, because $2 \times 5 = 10$.

   (B) No, $2\sqrt{5} = \sqrt{20}$, not $\sqrt{10}$.

   (C) Yes, multiplication distributes into the square root.

   (D) No, $2\sqrt{5} = 4\sqrt{5}$.

4. *How long will it take for $400 to earn $60 in simple interest at 5% per year?*

   Your Answer:

5. *What is the simplified form of $3^4 \cdot 3^2$?*

   (A) 3^6

   (B) 3^8

   (C) 9^6

   (D) 3^2

6. *What is 6.1×10^3 written in standard form?*

   (A) 61,000

   (B) 610

   (C) 6,100

   (D) 0.0061

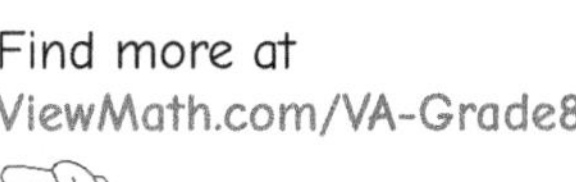
Find more at
ViewMath.com/VA-Grade8

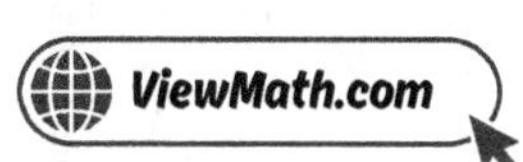

7. A bacterium has a length of 2×10^{-6} m. If 1 micrometer (μm) $= 10^{-6}$ m, what is the bacterium's length in micrometers?

- (A) $2 \times 10^{-12}\ \mu m$
- (B) $0.002\ \mu m$
- (C) $2\ \mu m$
- (D) $2,000\ \mu m$

8. Taxi A charges \$3 per mile. Taxi B charges \$36 for a 9-mile trip. Which taxi has the higher unit rate?

Your Answer:

9. Solve $\frac{2x-1}{3} = 5$.

- (A) $x = 7$
- (B) $x = 7.5$
- (C) $x = 8$
- (D) $x = 16$

10. Two numbers add to 25 and differ by 7. What is the larger number?

- (A) 14
- (B) 15
- (C) 16
- (D) 17

11. Function M: $y = 7x + 2$. Function N passes through $(0, 10)$ and $(5, 35)$. At $x = 0$, which function has a greater value?

- (A) Function M
- (B) Function N
- (C) They are equal at $x = 0$.
- (D) Cannot be determined.

12. A function table shows $(0, 2)$, $(1, 6)$, $(2, 10)$, $(3, 14)$. What is the constant rate of change?

Your Answer:

Find more at
ViewMath.com/VA-Grade8

ViewMath.com

13. The table below represents a linear function. What is the equation?

x	0	1	2	3	4
y	7	10	13	16	19

(A) $y = 7x + 3$

(B) $y = 3x + 10$

(C) $y = 3x + 7$

(D) $y = 10x + 7$

14. A student fills a glass with water, drinks half, then refills it. Describe the graph of water level vs. time.

Your Answer:

15. A figure is reflected over the y-axis. What happens to point $(7, -2)$?

(A) $(7, 2)$

(B) $(-7, 2)$

(C) $(-7, -2)$

(D) $(-2, 7)$

16. The two triangles shown below have matching angle measures. Are they congruent?

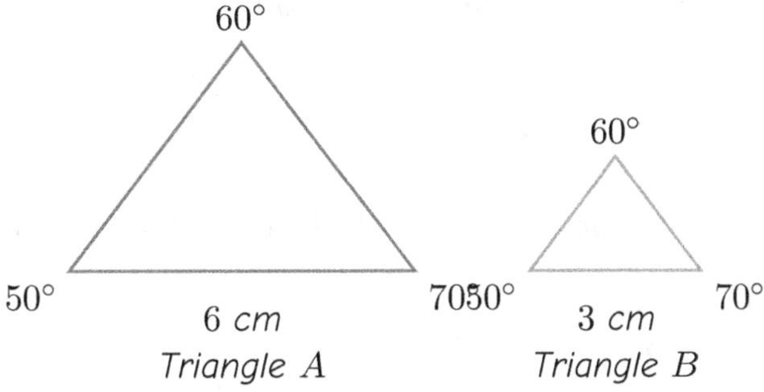

(A) Yes, because all three angles are equal.

(B) Yes, because one is a rotation of the other.

(C) No, because their side lengths are different.

(D) No, because their angles are in different positions.

17. Point (a, b) is rotated $180°$ and then reflected over the x-axis. The final image is:

(A) (a, b)

(B) $(-a, b)$

(C) $(a, -b)$

(D) $(-a, -b)$

18. A photo is 4 inches by 6 inches. It is enlarged by a scale factor of 2.5. What are the new dimensions?

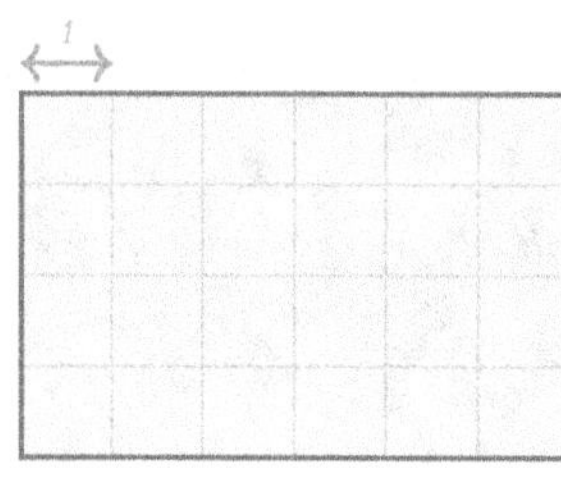

(A) 6.5 in by 8.5 in

(B) 8 in by 12 in

(C) 10 in by 15 in

(D) 10 in by 12 in

19. Corresponding angles formed by a transversal and two parallel lines are located:

(A) On opposite sides of the transversal, between the parallel lines

(B) On the same side of the transversal, in matching positions

(C) On opposite sides of the transversal, outside the parallel lines

(D) Adjacent to each other at one intersection

20. A ladder 15 ft long leans against a wall. The base is 9 ft from the wall. How high up the wall does the ladder reach?

(A) 6 ft

(B) 12 ft

(C) $\sqrt{306}$ ft

(D) 24 ft

Find more at
ViewMath.com/VA-Grade8

ViewMath.com

21. A cylinder has a volume of 450π cm^3. A cone has the same base and height. What is the cone's volume?

(A) 150π cm^3

(B) 225π cm^3

(C) 1350π cm^3

(D) 900π cm^3

22. Which statement about vertical angles is true?

(A) They are always supplementary.

(B) They are always complementary.

(C) They are always equal.

(D) They are always adjacent.

23. Find the total surface area of the cylinder. (Use $\pi \approx 3.14$.)

$r = 3\ cm$

12 cm

(A) $282.6\ cm^2$

(B) $226.08\ cm^2$

(C) $339.12\ cm^2$

(D) $113.04\ cm^2$

24. A pyramid has a pentagonal base with area $35\ cm^2$ and height $6\ cm$. What is the volume?

Your Answer:

25. A rectangle is 12×8 cm. A square of side 3 cm is cut from one corner. What is the remaining area?

(A) $96\ cm^2$

(B) $87\ cm^2$

(C) $93\ cm^2$

(D) $84\ cm^2$

Find more at
ViewMath.com/VA-Grade8

26. A scatter plot shows data points that form a curved (U-shaped) pattern. This is best described as:

(A) positive linear association

(B) negative linear association

(C) nonlinear association

(D) no association

27. A trend line has equation $y = 1.5x + 2$. Using this, what is y when $x = 4$?

(A) 6

(B) 7

(C) 8

(D) 10

28. Data was collected for x-values from 1 to 10. Using the model to predict y when $x = 50$ is an example of:

(A) interpolation

(B) extrapolation

(C) correlation

(D) association

29. 80 students were surveyed about playing sports and playing a musical instrument.

	Plays sport	No sport	Total
Plays instrument	15	25	40
No instrument	30	10	40
Total	45	35	80

Is there an association between playing a sport and playing an instrument? Justify using conditional relative frequencies.

Your Answer:

30. A jar has 6 green and 4 yellow balls. One ball is drawn and replaced. Then another is drawn. $P(\text{both green}) = ?$

(A) $\frac{36}{100}$

(B) $\frac{30}{90}$

(C) $\frac{6}{10}$

(D) $\frac{12}{20}$

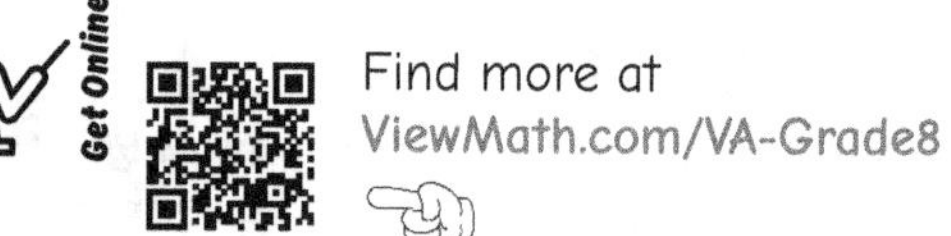

Find more at
ViewMath.com/VA-Grade8

 End of Practice Test 4 

Great job finishing the test!

☑ My Score

I got _____________ out of 30 questions right.

*Check your answers in the **Answer Key** at the back of the book.*

💡 *Review any questions you missed. That's how we learn!*

📊 Check Your Score Online!

Visit **ViewMath Academy** to enter your answers and see which topics you need to review. You can also explore lessons, take quizzes, track your scores, and save your progress!

viewmath.com/score/8.1.VA.19

Or go to viewmath.com/score and enter code: 8.1.VA.19

5

Practice Test 5

☑ 30 Questions

✏ Before You Start ✏

✓ **Read each question carefully** before choosing your answer.

✓ **Show your work** on scratch paper when you need to.

✓ **Skip hard questions** and come back to them later.

✓ **Check your answers** when you're done.

✓ **Take your time** — there's no rush!

⭐ You've Got This! ⭐

Do your best and show what you know!

1. A calculator shows $\sqrt{144} = 12$. Is $\sqrt{144}$ rational or irrational? Explain.

Your Answer

2. To convert $0.\overline{63}$ to a fraction, which equation should you set up after letting $x = 0.\overline{63}$?

(A) $10x - x = 63$

(B) $100x - x = 63$

(C) $100x - x = 6.3$

(D) $1000x - x = 63$

3. If the side length of a square is $\sqrt{50}$ cm, what is the perimeter?

(A) $4\sqrt{50} \approx 28.3$ cm

(B) 50 cm

(C) $\sqrt{200} \approx 14.1$ cm

(D) 200 cm

4. Look at the receipt below. The sales tax rate is 8%. What should the total be?

Item	Price
T-shirt	$20.00
Jeans	$45.00
Subtotal	$65.00
Tax (8%)	?
Total	?

(A) $Tax = \$5.00$; $Total = \$70.00$

(B) $Tax = \$5.20$; $Total = \$70.20$

(C) $Tax = \$6.50$; $Total = \$71.50$

(D) $Tax = \$8.00$; $Total = \$73.00$

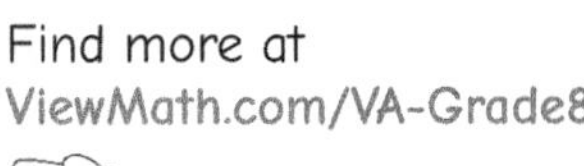

Find more at
ViewMath.com/VA-Grade8

5. *Simplify* $(2^3)^4$.

(A) 2^7

(B) 2^{12}

(C) 8^4

(D) 2^{64}

6. *A hydrogen atom has a radius of about* 2.5×10^{-11} *meters. How many places do you move the decimal to write this in standard form?*

Your Answer:

7. *What is* $5.2 \times 10^6 + 3.8 \times 10^6$?

(A) 9×10^6

(B) 9×10^{12}

(C) 19.76×10^6

(D) 9×10^{36}

8. *Which ordered pair could NOT lie on the graph of a proportional relationship?*

(A) $(0,0)$

(B) $(1,5)$

(C) $(3,15)$

(D) $(2,13)$

9. *Solve* $8 - 3x = 2$.

(A) $x = -2$

(B) $x = 2$

(C) $x = 3$

(D) $x = \frac{10}{3}$

10. *A rectangle has a perimeter of 48 m. Its length is twice its width. Find the dimensions.*

Your Answer:

Find more at
ViewMath.com/VA-Grade8

ViewMath.com

11. Function A: $y = 6x + 4$. Function B passes through $(0, 4)$ and $(3, 19)$. What is the rate of change of Function B?

Your Answer:

12. Which equation is nonlinear?

(A) $y = -5x + 2$

(B) $y = 10$

(C) $y = 3x$

(D) $y = x^2 - 1$

13. Find the equation of the line through $(1, 3)$ and $(4, 12)$.

Your Answer:

14. A car accelerates from a stop, cruises at constant speed, then brakes. Which describes the speed graph?

(A) Constant, increasing, decreasing

(B) Increasing, constant, decreasing

(C) Increasing, decreasing, constant

(D) Decreasing, constant, increasing

15. Point $P(2, 0)$ is rotated $90°$ counterclockwise around the origin. Where does P' land?

(A) $(0, -2)$

(B) $(0, 2)$

(C) $(-2, 0)$

(D) $(2, 0)$

16. $\triangle PQR \cong \triangle STU$. If $\angle P = 72°$ and $\angle Q = 53°$, what is $\angle U$?

Your Answer:

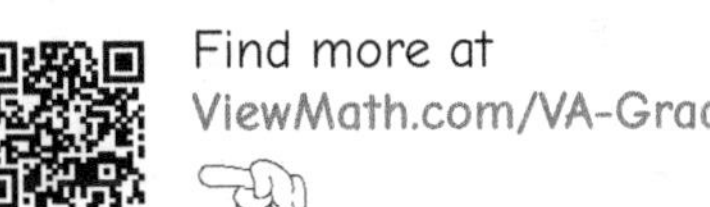
Find more at
ViewMath.com/VA-Grade8

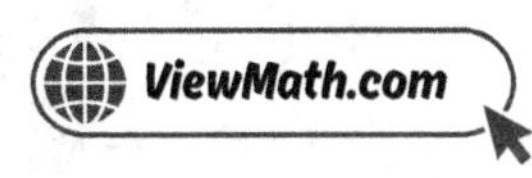

17. A point is dilated by factor 3 from the origin. If the image is $(9, -15)$, what was the original point?

    (A) $(27, -45)$

    (B) $(3, -5)$

    (C) $(6, -12)$

    (D) $(12, -18)$

18. A dilation with scale factor $k = 3$ makes a figure:

    (A) 3 times smaller

    (B) 3 times larger

    (C) The same size

    (D) 9 times larger

19. In an isosceles triangle, the two base angles are each $54°$. What is the vertex angle?

    (A) $54°$

    (B) $72°$

    (C) $108°$

    (D) $126°$

20. A right triangle has legs 1 and 1. What is the exact length of the hypotenuse?

    (A) 2

    (B) $\sqrt{2}$

    (C) 1

    (D) $\sqrt{3}$

21. A cone has volume 100 cm^3. A cylinder with the same base and height has volume ___ cm^3.

    Your Answer:

22. Two adjacent angles on a straight line measure $(3x)°$ and $(x + 20)°$. What is x?

    (A) 20

    (B) 40

    (C) 45

    (D) 50

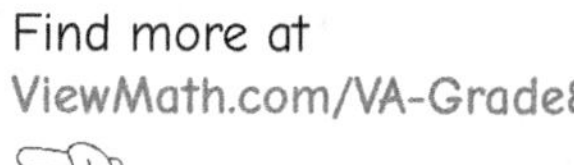
Find more at
ViewMath.com/VA-Grade8

ViewMath.com

23. A rectangular prism has faces with areas 15 cm^2, 20 cm^2, and 12 cm^2. What is the total surface area?

(A) 47 cm^2

(B) 94 cm^2

(C) 188 cm^2

(D) 60 cm^2

24. What is the volume of a square pyramid with base side 6 cm and height 10 cm?

(A) 360 cm^3

(B) 120 cm^3

(C) 180 cm^3

(D) 60 cm^3

25. A circle with radius 6 cm has a square with side 6 cm cut from its interior. What is the remaining area? (Use $\pi \approx 3.14$.)

(A) 77.04 cm^2

(B) 113.04 cm^2

(C) 36 cm^2

(D) 149.04 cm^2

26. Which variable typically goes on the x-axis in a scatter plot?

(A) The dependent variable

(B) The response variable

(C) The independent (explanatory) variable

(D) The variable with the largest values

27. A trend line passes through $(2, 10)$ and $(8, 4)$. Find the slope and y-intercept.

Your Answer:

28. Using the model $y = 4x + 7$, for what value of x is $y = 35$?

(A) 7

(B) 6

(C) 8

(D) 10

Find more at
ViewMath.com/VA-Grade8

29. Complete the two-way table and find the relative frequencies for each cell (as fractions of the total).

	Homework Yes	Homework No	Total
Grade A	24	6	
Not Grade A	16	14	
Total			

Your Answer

30. $P(A) = 0.6$, $P(B) = 0.4$. If A and B are independent, $P(A \text{ and } B) =?$

(A) 1.0

(B) 0.24

(C) 0.20

(D) 0.10

 *End of Practice Test 5* 

Great job finishing the test!

☑ My Score

I got ___________ out of 30 questions right.

*Check your answers in the **Answer Key** at the back of the book.*

💡 *Review any questions you missed. That's how we learn!*

📊 Check Your Score Online!

Visit **ViewMath Academy** to enter your answers and see which topics you need to review. You can also explore lessons, take quizzes, track your scores, and save your progress!

viewmath.com/score/8.1.VA.20

Or go to *viewmath.com/score* and enter code: 8.1.VA.20

Practice Test 6

 30 Questions

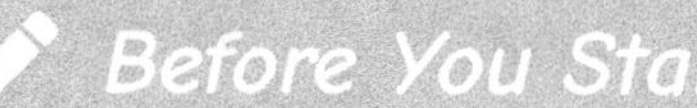
✏ Before You Start ✏

- ✔ **Read each question carefully** before choosing your answer.
- ✔ **Show your work** on scratch paper when you need to.
- ✔ **Skip hard questions** and come back to them later.
- ✔ **Check your answers** when you're done.
- ✔ **Take your time** — there's no rush!

 You've Got This! 

Do your best and show what you know!

1. Is $\frac{5}{6}$ rational or irrational? Explain.

Your Answer:

2. When converting $0.\overline{123}$ to a fraction, by what power of 10 should you multiply both sides?

(A) 10

(B) 100

(C) 1000

(D) 10000

3. Order from least to greatest: π, $\sqrt{10}$, 3.2.

(A) π, 3.2, $\sqrt{10}$

(B) 3.2, π, $\sqrt{10}$

(C) $\sqrt{10}$, π, 3.2

(D) π, $\sqrt{10}$, 3.2

4. Maria deposits \$1,000 in a savings account that pays 3% simple interest per year. How much will she have after 4 years?

(A) \$1,030

(B) \$1,120

(C) \$1,300

(D) \$1,012

5. Which of the following is equal to $\frac{8^6}{8^6}$?

(A) 8^{36}

(B) 8^0

(C) 0

(D) 8

6. What is 9.03×10^{-5} written in standard form?

(A) 903,000

(B) 0.0000903

(C) 0.000903

(D) 0.00903

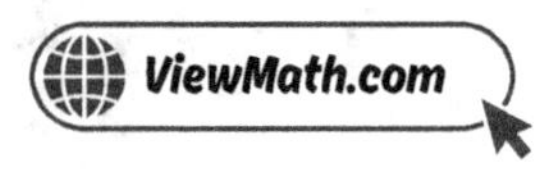

7. A light year is approximately 9.5×10^{12} km. If a star is 4 light years away, how far is it in kilometers?

(A) 3.8×10^{13} km

(B) 3.8×10^{12} km

(C) 13.5×10^{12} km

(D) 9.5×10^{48} km

8. A graph shows a straight line through $(0,0)$ and $(2,9)$. What is the unit rate?

(A) 2

(B) 4.5

(C) 9

(D) 18

9. Solve $2(x+5) = 3x - 1$.

(A) $x = 9$

(B) $x = 10$

(C) $x = 11$

(D) $x = 12$

10. Maria has 15 coins, all nickels and dimes. The coins are worth \$1.10. How many dimes does she have?

(A) 5

(B) 6

(C) 7

(D) 8

11. Function G is shown in the table:

x	0	1	2	3	4
y	2	5	8	11	14

Function H: $y = 4x + 1$. Which function has the greater value at $x = 4$?

(A) Function G

(B) Function H

(C) They are equal.

(D) Cannot be determined.

12. Which graph represents a linear function?

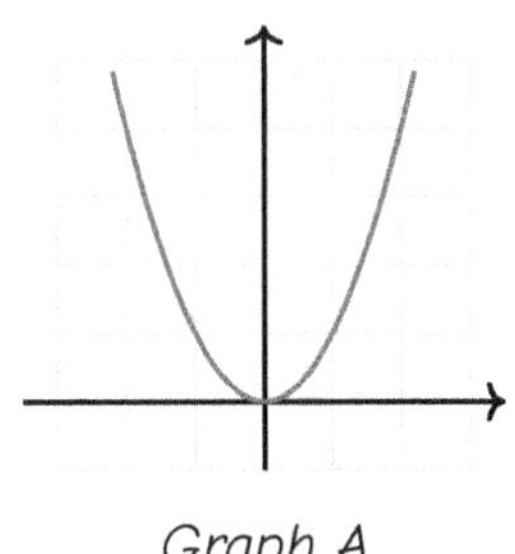

Graph A

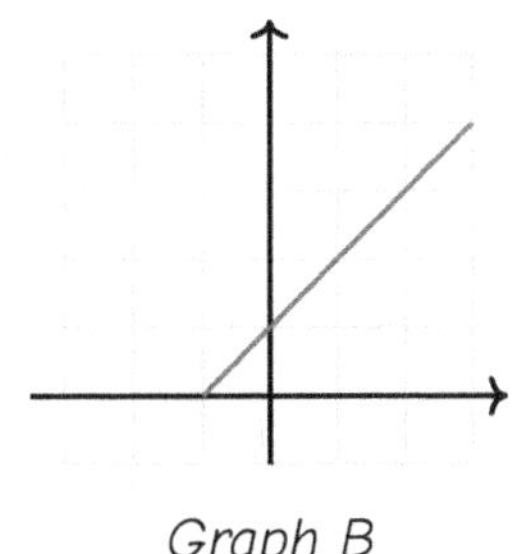

Graph B

(A) Graph A

(B) Graph B

(C) Both

(D) Neither

13. A function has a slope of -3 and passes through $(2, 1)$. What is the y-intercept?

Your Answer:

14. Which scenario matches a graph that is decreasing and linear?

(A) A bank account earning interest

(B) A car driving at constant speed losing fuel at a steady rate

(C) A ball bouncing up and down

(D) A tree growing over many years

15. After a translation, a triangle's longest side was originally 8 cm. What is the longest side of the image?

(A) 4 cm

(B) 8 cm

(C) 16 cm

(D) It depends on the direction of the translation.

Find more at
ViewMath.com/VA-Grade8

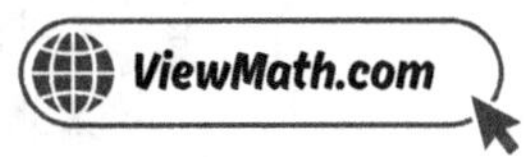

16. A student says two circles with radius 5 cm are always congruent. Is this correct?

(A) No, circles cannot be congruent.

(B) No, they need the same center too.

(C) Yes, because they have the same radius.

(D) Yes, but only if they are in the same position.

17. Translate the point $(4, -6)$ by $(-3, 8)$. What is the image?

Your Answer

18. True or false: All rectangles are similar to each other.

Your Answer

19. A triangle has angles $x°$, $2x°$, and $3x°$. What is the value of x?

(A) 60

(B) 30

(C) 36

(D) 45

20. A right triangle has legs a and $b = 4a$. If the hypotenuse is $\sqrt{17} \cdot a$, verify using the Pythagorean Theorem.

Your Answer

21. A cone has radius 3 m and height 12 m. What is its volume in terms of π?

(A) $108\pi \ m^3$

(B) $36\pi \ m^3$

(C) $12\pi \ m^3$

(D) $324\pi \ m^3$

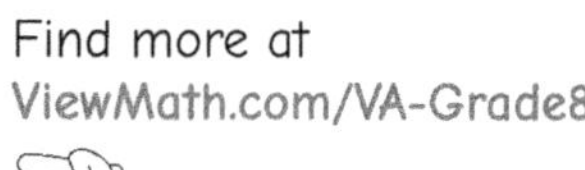
Find more at
ViewMath.com/VA-Grade8

22. Two vertical angles measure $(4x + 8)°$ and $(6x - 12)°$. Find x.

Your Answer:

23. A triangular pyramid (tetrahedron) has 4 equilateral triangular faces, each with area 25 cm². What is the total surface area?

(A) 50 cm²

(B) 75 cm²

(C) 100 cm²

(D) 125 cm²

24. A triangular pyramid has a base triangle with area 30 cm² and a height of 9 cm. What is the volume?

(A) 270 cm³

(B) 90 cm³

(C) 135 cm³

(D) 45 cm³

25. A composite figure is made of a semicircle (diameter 8 cm) on top of a rectangle (8 × 5 cm). What is the total area? (Use $\pi \approx 3.14$.)

(A) 65.12 cm²

(B) 40 cm²

(C) 90.24 cm²

(D) 50.24 cm²

26. Data: $(1, 2), (2, 4), (3, 5), (4, 8), (5, 9)$. The dots appear to go upward from left to right. The association is:

(A) negative

(B) positive

(C) no association

(D) nonlinear

Find more at
ViewMath.com/VA-Grade8

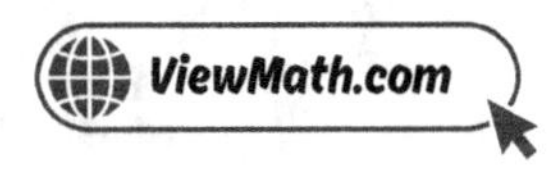

27. *Data:* $(1,3), (2,5), (3,4), (4,6), (5,8)$. *A trend line through* $(1,3)$ *and* $(5,8)$ *has slope:*

(A) $\frac{5}{4}$

(B) $\frac{4}{5}$

(C) 1

(D) 2

28. What is extrapolation?

(A) Making predictions within the range of the data

(B) Making predictions far beyond the range of the data

(C) Removing outliers from the data

(D) Finding the exact equation of the line

29. A survey asks students about their favorite sport (soccer or basketball) and grade (7th or 8th). 20 7th-graders chose soccer. This 20 is a:

(A) marginal frequency

(B) joint frequency

(C) relative frequency

(D) total frequency

30. What is the probability of an impossible event?

(A) 1

(B) 0.5

(C) 0

(D) -1

 # End of Practice Test 6 

Great job finishing the test!

 My Score

I got _____________ out of 30 questions right.

Check your answers in the **Answer Key** at the back of the book.

💡 Review any questions you missed. That's how we learn!

📊 Check Your Score Online!

Visit **ViewMath Academy** to enter your answers and see which topics you need to review. You can also explore lessons, take quizzes, track your scores, and save your progress!

viewmath.com/score/8.1.VA.21

Or go to viewmath.com/score and enter code: 8.1.VA.21

Practice Test 7

📋 30 Questions

✏️ Before You Start ✏️

- ✔ **Read each question carefully** before choosing your answer.
- ✔ **Show your work** on scratch paper when you need to.
- ✔ **Skip hard questions** and come back to them later.
- ✔ **Check your answers** when you're done.
- ✔ **Take your time** — there's no rush!

⭐ You've Got This! ⭐

Do your best and show what you know!

1. Give an example of an irrational number between 1 and 2.

Your Answer:

2. A student converts $0.\overline{36}$ and gets $\frac{36}{100}$. What mistake did the student make?

(A) The student divided by 100 instead of 99.

(B) The student forgot to simplify.

(C) The student treated a repeating decimal as a terminating decimal.

(D) The student multiplied by 10 instead of 100.

3. Estimate $\sqrt{3} + \sqrt{7}$ to one decimal place.

Your Answer:

4. An investment of \$800 earns \$96 in simple interest over 4 years. What is the annual interest rate?

(A) 2%

(B) 3%

(C) 4%

(D) 12%

5. Write $\frac{7^3 \cdot 7^4}{7^5}$ as a single power of 7.

Your Answer:

6. Which of the following is 350,000 written in scientific notation?

(A) 35×10^4

(B) 3.5×10^5

(C) 3.5×10^4

(D) 0.35×10^6

7. Earth's mass is about 6×10^{24} kg and Jupiter's mass is about 1.9×10^{27} kg. About how many times more massive is Jupiter than Earth? Round to the nearest whole number.

Your Answer:

8. Which table does NOT represent a proportional relationship?

A)

x	1	2
y	4	8

B)

x	2	4
y	5	10

C)

x	3	6
y	9	15

D)

x	5	10
y	15	30

9. Solve $3x + 7 = 22$.

A) $x = 3$ B) $x = 5$

C) $x = 7$ D) $x = 10$

10. A rectangle's perimeter is 34 cm. The length is 5 cm more than the width. What is the width?

A) 4 cm   B) 5 cm

C) 6 cm   D) 7 cm

11. Function M has slope 4 and y-intercept 10. Function N has slope 4 and y-intercept 3. Are their graphs parallel? Write Yes or No.

Your Answer:

Find more at
ViewMath.com/VA-Grade8

ViewMath.com

12. A student says $y = 3x^2$ is linear because the coefficient is 3. What is wrong with this reasoning?

(A) The coefficient should be 1 for it to be linear.

(B) The variable x is raised to the 2nd power, making it nonlinear.

(C) The function has no y-intercept.

(D) Nothing is wrong; it is linear.

13. A function has slope -5 and passes through $(0, 30)$. What is its equation?

(A) $y = 30x - 5$

(B) $y = -5x - 30$

(C) $y = -5x + 30$

(D) $y = 5x + 30$

14. A roller coaster goes up, down sharply, up again, then levels off. How many sections does the graph have?

(A) 2

(B) 3

(C) 4

(D) 5

15. Point $M(5, -3)$ is reflected over the y-axis. What are the coordinates of M'?

Your Answer:

16. Triangle A has sides 5, 7, and 9. Triangle B has sides 5, 7, and 9. What can you conclude?

(A) The triangles are similar but not congruent.

(B) The triangles are congruent.

(C) The triangles have different angles.

(D) Not enough information to decide.

17. What is the image of $(5, -3)$ after a translation of $(-2, 4)$?

(A) $(7, -7)$

(B) $(3, 1)$

(C) $(3, -7)$

(D) $(7, 1)$

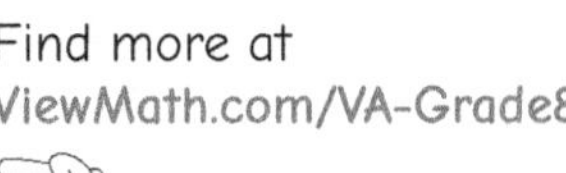

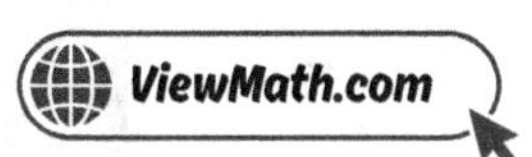

18. Two similar rectangles have corresponding sides in the ratio 2 : 5. If the shorter side of the smaller rectangle is 8 cm, what is the shorter side of the larger rectangle?

(A) 3.2 cm

(B) 13 cm

(C) 16 cm

(D) 20 cm

19. A right triangle has one angle of 32°. What is the third angle?

(A) 148°

(B) 58°

(C) 68°

(D) 48°

20. A right triangle has legs 10 and 24. What is the hypotenuse?

Your Answer

21. What is the volume of a cylinder with radius 2 cm and height 7 cm in terms of π?

(A) $14\pi \ cm^3$

(B) $28\pi \ cm^3$

(C) $56\pi \ cm^3$

(D) $7\pi \ cm^3$

22. Two supplementary angles are $(2x + 15)°$ and $(3x - 10)°$. What is x?

(A) 25

(B) 35

(C) 37

(D) 55

23. Find the surface area of a rectangular prism with length 7 cm, width 4 cm, and height 2 cm.

Your Answer

Find more at
ViewMath.com/VA-Grade8

24. A rectangular pyramid is shown. What is the volume?

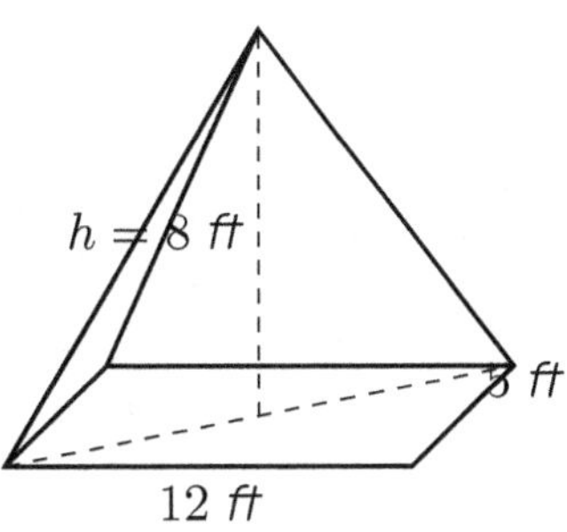

(A) 480 ft^3

(B) 160 ft^3

(C) 240 ft^3

(D) 320 ft^3

25. A cross shape is formed by overlapping two rectangles (12×4 cm each) at their centers. The overlap is a 4×4 cm square. What is the total area of the cross?

Your Answer:

26. A scatter plot has a strong upward trend. Which r-value (correlation) is most likely?

(A) $r = -0.9$

(B) $r = 0.1$

(C) $r = 0.9$

(D) $r = 0$

27. *Data:* $(1, 8), (2, 7), (3, 5), (4, 4), (5, 2)$. *Draw a trend line and write its equation.*

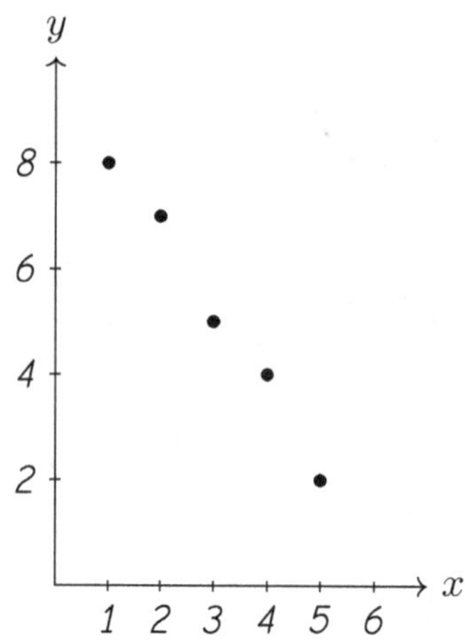

Your Answer:

28. *A linear model is* $y = 2x + 5$. *What does the slope represent?*

(A) The starting value

(B) The rate of change per unit increase in x

(C) The x-intercept

(D) The total value of y

29. *Using the table above, what proportion of pizza lovers are boys?*

(A) $\frac{18}{42}$

(B) $\frac{24}{42}$

(C) $\frac{24}{80}$

(D) $\frac{24}{40}$

30. *A card is drawn from a standard 52-card deck. What is* $P(heart)$?

(A) $\frac{1}{52}$

(B) $\frac{1}{13}$

(C) $\frac{1}{4}$

(D) $\frac{1}{2}$

 # End of Practice Test 7 

Great job finishing the test!

My Score

I got _____________ out of 30 questions right.

*Check your answers in the **Answer Key** at the back of the book.*

💡 *Review any questions you missed. That's how we learn!*

📊 Check Your Score Online!

Visit **ViewMath Academy** to enter your answers and see which topics you need to review. You can also explore lessons, take quizzes, track your scores, and save your progress!

viewmath.com/score/8.1.VA.22

Or go to viewmath.com/score and enter code: 8.1.VA.22

Practice Test 8

📋 30 Questions

✏️ Before You Start ✏️

✔ **Read each question carefully** before choosing your answer.

✔ **Show your work** on scratch paper when you need to.

✔ **Skip hard questions** and come back to them later.

✔ **Check your answers** when you're done.

✔ **Take your time** — there's no rush!

⭐ You've Got This! ⭐

Do your best and show what you know!

1. Which of the following is a true statement?

   (A) Every square root is irrational.

   (B) Every integer is irrational.

   (C) Every integer is rational.

   (D) Every decimal is irrational.

2. The table below shows repeating decimals and their fraction equivalents. Find the missing fraction for $0.\overline{81}$.

Decimal	Unsimplified Fraction	Simplest Form
$0.\overline{27}$	$\frac{27}{99}$	$\frac{3}{11}$
$0.\overline{54}$	$\frac{54}{99}$	$\frac{6}{11}$
$0.\overline{81}$	?	?

Your Answer:

3. A square patio has an area of 75 square feet. Estimate the side length to one decimal place.

Your Answer:

4. Sara earned \$45 in simple interest on an investment at 3% per year for 5 years. What was the principal?

   (A) \$300

   (B) \$675

   (C) \$150

   (D) \$900

5. Evaluate $5^0 + 5^{-1}$. Express your answer as a decimal.

Your Answer:

Find more at
ViewMath.com/VA-Grade8

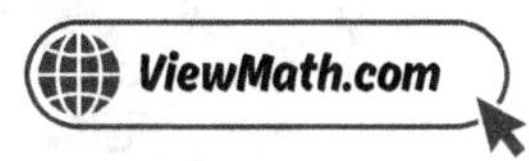

6. The bar chart compares the masses of four animals. Which animal's mass is closest to 1×10^2 kilograms?

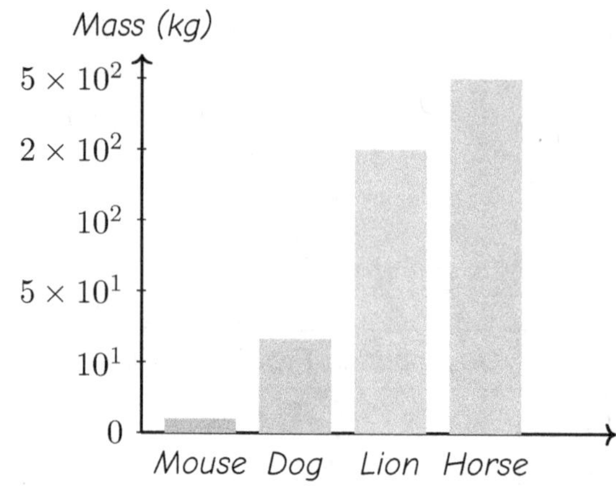

(A) Mouse

(B) Dog

(C) Lion

(D) Horse

7. A phone stores 6.4×10^{10} bytes. A photo uses 4×10^6 bytes. How many photos can the phone store?

(A) 1.6×10^3

(B) 1.6×10^4

(C) 1.6×10^{16}

(D) 2.4×10^4

8. Machine A fills $y = 12x$ bottles per hour. Machine B fills 50 bottles in 5 hours. Which machine is faster?

(A) Machine A

(B) Machine B

(C) They fill at the same rate.

(D) Cannot be determined.

9. When solving an equation, the variable disappears and you get $5 = 5$. This means:

(A) The equation has no solution.

(B) The equation has exactly one solution, $x = 5$.

(C) The equation has infinitely many solutions.

(D) You made a mistake.

10. One number is 4 more than another. Their sum is 36. Find both numbers.

Your Answer

11. The table shows Function P. Function Q is $y = 2x + 9$. Which function has the greater initial value?

x	0	1	2	3
$P(x)$	5	8	11	14

(A) Function P

(B) Function Q

(C) They have the same initial value.

(D) Cannot be determined.

12. The function $y = -2x + 6$ is:

(A) Nonlinear, because the slope is negative.

(B) Nonlinear, because y can be negative.

(C) Linear, because it has the form $y = mx + b$.

(D) Linear, because it passes through the origin.

13. The table shows $(2, 9)$ and $(6, 25)$. Find the slope of the line.

Your Answer

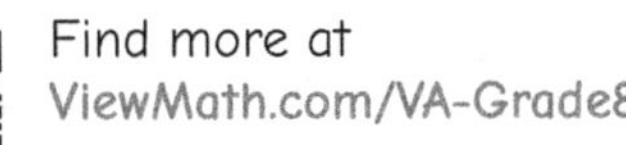
Find more at
ViewMath.com/VA-Grade8

ViewMath.com

14. The graph below shows a runner's distance from the start over time. Write a short story that matches the graph.

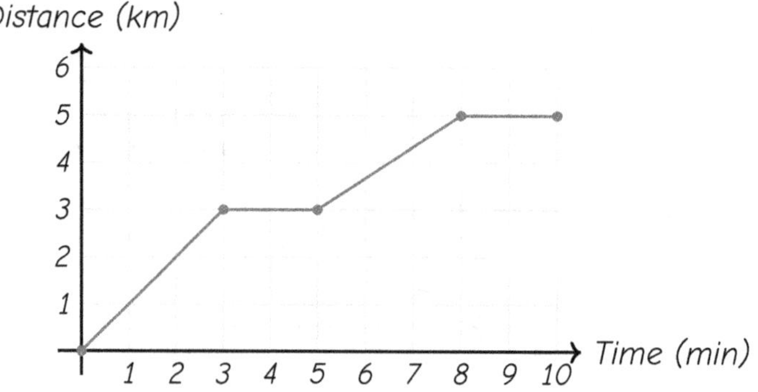

Your Answer:

15. Name the three rigid transformations.

Your Answer:

16. $\triangle ABC \cong \triangle DEF$. The perimeter of $\triangle ABC$ is 30 cm. What is the perimeter of $\triangle DEF$?

Your Answer:

17. Point $(5, 2)$ is rotated $180°$ around the origin. What are the coordinates of the image?

Your Answer:

18. Are all squares similar to each other?

(A) No, because they can have different side lengths.

(B) No, because they can have different angles.

(C) Yes, because all squares have equal angles and proportional sides.

(D) Yes, but only if they have the same perimeter.

19. In a triangle, the angles are $(2x + 5)°$, $(3x)°$, and $(x + 25)°$. Find x.

Your Answer:

20. Is the triangle with the given side lengths a right triangle?

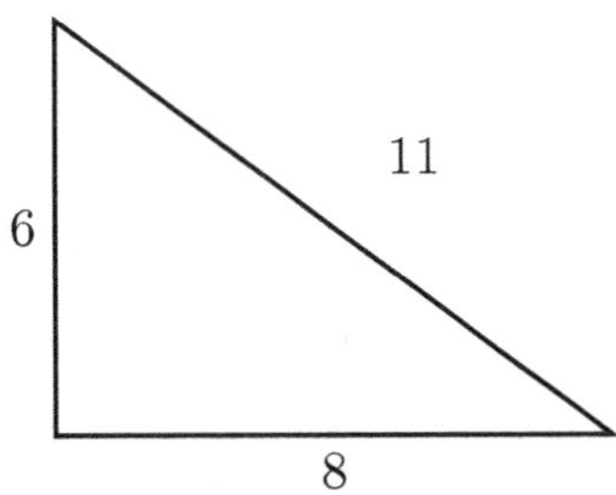

(A) Yes, because $6^2 + 8^2 = 11^2$

(B) Yes, because $6 + 8 > 11$

(C) No, because $6^2 + 8^2 \neq 11^2$

(D) No, because the triangle has no right angle symbol

21. A cylindrical can has radius 3 in and height 8 in. How much soup can it hold? Use $\pi \approx 3.14$.

Your Answer:

22. Two angles are supplementary. One angle is 125°. What is the other?

(A) 55°

(B) 35°

(C) 125°

(D) 235°

23. Find the lateral surface area of a cylinder with radius 7 cm and height 10 cm. (Use $\pi \approx 3.14$.)

(A) 439.6 cm^2

(B) 219.8 cm^2

(C) 307.72 cm^2

(D) 747.32 cm^2

24. A square pyramid has volume 48 in^3 and height 4 in. What is the base side length?

Your Answer:

25. Find the perimeter of a figure that is a 6×6 cm square with one side replaced by a semicircle (diameter 6 cm). (Use $\pi \approx 3.14$.)

Your Answer:

26. In a scatter plot, several data points bunch together near $(3, 20)$. This grouping is called:

(A) an outlier

(B) a gap

(C) a cluster

(D) a negative association

27. Data: $(1, 2), (2, 4), (3, 7), (4, 8), (5, 11)$. Estimate the slope of a reasonable trend line.

Your Answer:

Find more at
ViewMath.com/VA-Grade8

ViewMath.com

28. In the model $y = 6x + 12$, if x increases by 3 units, how much does y increase?

(A) 3

(B) 6

(C) 18

(D) 30

29. Using the table from q19, what percentage of boys prefer reading? What percentage of girls prefer reading? Is there an association?

Your Answer:

30. $P(event) = \frac{3}{4}$. What is $P(not\ event)$?

(A) $\frac{3}{4}$

(B) $\frac{1}{4}$

(C) $\frac{1}{2}$

(D) 0

Find more at
ViewMath.com/VA-Grade8

 End of Practice Test 8 

Great job finishing the test!

☑ My Score

I got ___________ out of 30 questions right.

💡 Review any questions you missed. That's how we learn!

📊 Check Your Score Online!

Visit **ViewMath Academy** to enter your answers and see which topics you need to review. You can also explore lessons, take quizzes, track your scores, and save your progress!

viewmath.com/score/8.1.VA.23

Or go to viewmath.com/score and enter code: 8.1.VA.23

Practice Test 9

 30 Questions

✏️ Before You Start ✏️

- ✔ **Read each question carefully** before choosing your answer.
- ✔ **Show your work** on scratch paper when you need to.
- ✔ **Skip hard questions** and come back to them later.
- ✔ **Check your answers** when you're done.
- ✔ **Take your time** — there's no rush!

⭐ You've Got This! ⭐

Do your best and show what you know!

1. Which of the following could NOT be the decimal expansion of a rational number?

   (A) 2.500

   (B) $0.\overline{9}$

   (C) $1.41421356\ldots$ *(non-repeating)*

   (D) $0.\overline{36}$

2. What is $0.\overline{4}$ written as a fraction?

   (A) $\frac{4}{10}$

   (B) $\frac{4}{9}$

   (C) $\frac{2}{5}$

   (D) $\frac{1}{4}$

3. Between which two consecutive integers does $3\sqrt{2}$ lie?

   (A) 3 and 4

   (B) 4 and 5

   (C) 5 and 6

   (D) 6 and 7

4. A waiter earns a 15% tip on a \$48 meal. How much is the tip?

   (A) \$7.20

   (B) \$6.40

   (C) \$4.80

   (D) \$15.00

5. Which expression is equivalent to $10^{-4} \cdot 10^7$?

   (A) 10^{-28}

   (B) 10^{-3}

   (C) 10^3

   (D) 10^{11}

6. Which is larger: 8×10^5 or 3×10^6?

   (A) 8×10^5, because $8 > 3$

   (B) 3×10^6, because the exponent is larger

   (C) They are equal

   (D) It cannot be determined

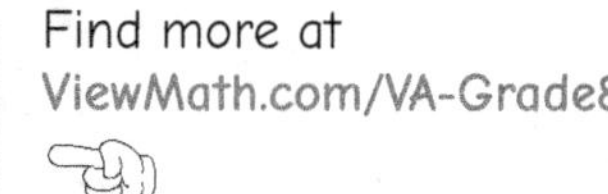
Find more at
ViewMath.com/VA-Grade8

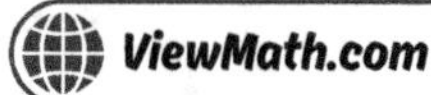

7. *Simplify $(7 \times 10^{-3})(8 \times 10^{-2})$. Write in proper scientific notation.*

(A) 56×10^{-5}

(B) 5.6×10^{-5}

(C) 5.6×10^{-4}

(D) 5.6×10^{6}

8. *Two delivery services are compared. Service P delivers $y = 4x$ packages per hour. Service Q's data is shown in the table.*

Hours (x)	2	4	6	8
Packages (y)	10	20	30	40

How many more packages does Service Q deliver than Service P in 10 hours?

Your Answer:

9. *Solve $4(x - 3) = 2(x + 1)$.*

(A) $x = 5$

(B) $x = 7$

(C) $x = 8$

(D) $x = 10$

10. *A store sells T-shirts for \$12 and hats for \$8. Use the information in the table to find how many of each were sold.*

	Value
Total items sold	15
Total revenue	\$148

Your Answer:

  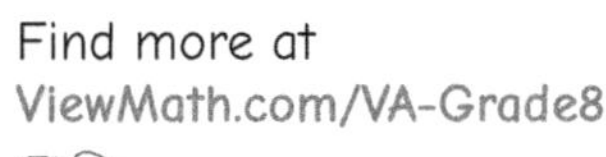
Find more at
ViewMath.com/VA-Grade8

11. Function R is shown in the table:

x	0	2	4	6
y	1	9	17	25

Function S: $y = 3x + 1$. Which has a greater initial value?

(A) Function R

(B) Function S

(C) They have the same initial value.

(D) Cannot be determined.

12. Which equation represents a linear function?

(A) $y = x^2 + 3$

(B) $y = \frac{1}{x}$

(C) $y = 4x - 7$

(D) $y = 2^x$

13. A line passes through $(3, 7)$ and $(6, 16)$. What is the y-intercept?

(A) -2

(B) 0

(C) 1

(D) 3

14. A flat (horizontal) section of a graph of height vs. time means:

(A) The height is increasing.

(B) The height is decreasing.

(C) The height is staying the same.

(D) Time has stopped.

15. A rectangle has vertices at $(1, 2)$, $(5, 2)$, $(5, 4)$, and $(1, 4)$. After a translation of 3 units right and 1 unit up, what are the coordinates of the vertex that was at $(1, 2)$?

(A) $(4, 3)$

(B) $(4, 1)$

(C) $(2, 5)$

(D) $(-2, 3)$

Find more at
ViewMath.com/VA-Grade8

ViewMath.com

16. Triangle $ABC \cong$ Triangle DEF. If $AB = 8$ cm, what is DE?

- (A) 4 cm
- (B) 8 cm
- (C) 16 cm
- (D) Cannot be determined

17. Which coordinate rule represents a reflection over the x-axis?

- (A) $(x, y) \to (-x, y)$
- (B) $(x, y) \to (x, -y)$
- (C) $(x, y) \to (-x, -y)$
- (D) $(x, y) \to (-y, x)$

18. A rectangle is 3 cm by 5 cm. Another rectangle is 6 cm by 9 cm. Are they similar?

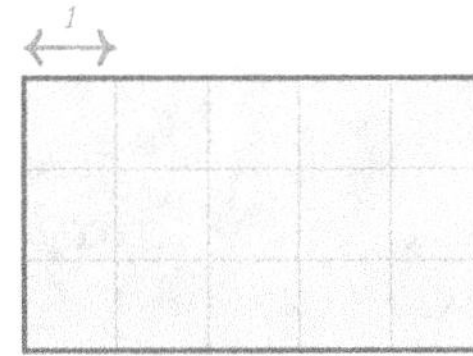

- (A) Yes, with scale factor 2
- (B) Yes, with scale factor 3
- (C) No, the sides are not proportional.
- (D) Yes, with scale factor $\frac{3}{5}$

19. An exterior angle of a triangle is $140°$. What is the adjacent interior angle?

- (A) $40°$
- (B) $140°$
- (C) $50°$
- (D) $220°$

20. An isosceles right triangle has legs of length 10. What is the hypotenuse?

- (A) 20
- (B) $10\sqrt{2}$
- (C) $\sqrt{10}$
- (D) 100

21. The volume of a sphere with radius r is $\frac{4}{3}\pi r^3$. If the radius is tripled, the new volume is:

(A) 3 times the original

(B) 9 times the original

(C) 27 times the original

(D) 81 times the original

22. Two angles are complementary. One angle is 37°. What is the other?

(A) 143°

(B) 53°

(C) 37°

(D) 63°

23. What is the surface area of a rectangular prism with length 12 m, width 5 m, and height 3 m?

(A) 222 m^2

(B) 180 m^2

(C) 111 m^2

(D) 162 m^2

24. A rectangular pyramid has a base of 8×5 m and a height of 12 m. What is its volume?

(A) 480 m^3

(B) 160 m^3

(C) 240 m^3

(D) 320 m^3

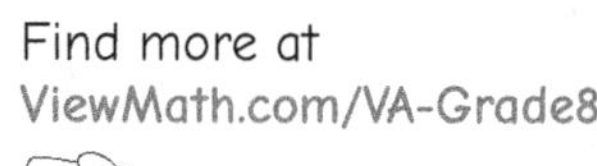

Find more at
ViewMath.com/VA-Grade8

25. *Find the area of the shaded composite figure.*

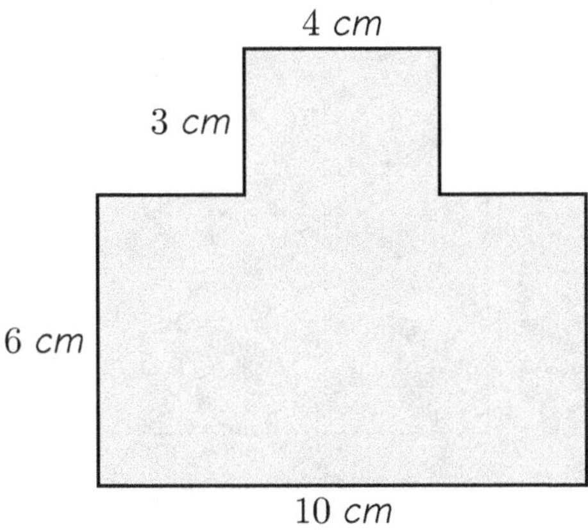

(A) 60 cm²   (B) 72 cm²

(C) 78 cm²   (D) 90 cm²

26. *True or false: A scatter plot can show both clustering and outliers at the same time.*

(A) *True — they describe different features of the data.*

(B) *False — data has either clusters or outliers, not both.*

(C) *True — but only if there is no association.*

(D) *False — outliers are always in clusters.*

27. A trend line is drawn through the scatter plot. What is the approximate slope of the line?

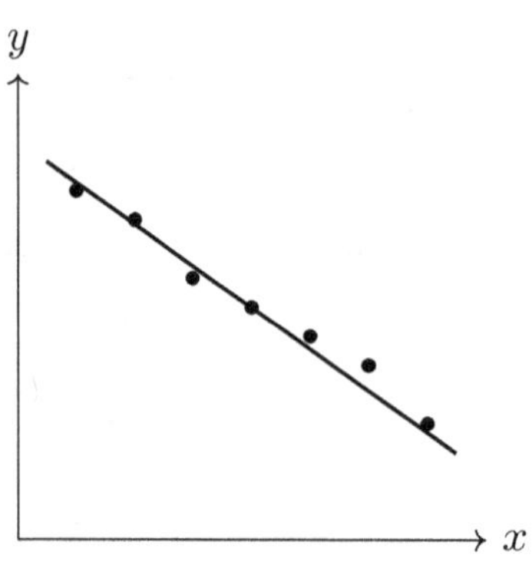

(A) $\approx -\frac{5}{7}$

(B) $\approx \frac{5}{7}$

(C) ≈ -2

(D) ≈ 1

28. A model gives $y = 5x + 20$ where x is weeks and y is savings (dollars). If $x = 8$, what is the predicted savings?

(A) $45

(B) $60

(C) $68

(D) $40

29. Which of these is NOT a question you can answer with a two-way table?

(A) What fraction of boys prefer soccer?

(B) Is there an association between grade and preferred lunch?

(C) What is the mean score on the math test?

(D) How many 8th graders chose art?

30. Explain the difference between independent and dependent events with an example.

Your Answer:

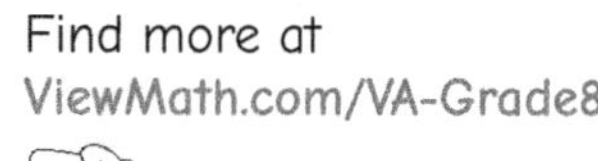
Find more at
ViewMath.com/VA-Grade8

 # End of Practice Test 9 

Great job finishing the test!

☑ My Score

I got _____________ out of 30 questions right.

Check your answers in the **Answer Key** at the back of the book.

💡 Review any questions you missed. That's how we learn!

📊 Check Your Score Online!

Visit **ViewMath Academy** to enter your answers and see which topics you need to review. You can also explore lessons, take quizzes, track your scores, and save your progress!

viewmath.com/score/8.1.VA.24

Or go to viewmath.com/score and enter code: 8.1.VA.24

Practice Test 10

30 Questions

✏️ Before You Start ✏️

- ✓ **Read each question carefully** before choosing your answer.
- ✓ **Show your work** on scratch paper when you need to.
- ✓ **Skip hard questions** and come back to them later.
- ✓ **Check your answers** when you're done.
- ✓ **Take your time** — there's no rush!

⭐ You've Got This! ⭐

Do your best and show what you know!

1. Look at the Venn diagram below. In which region does $\sqrt{5}$ belong?

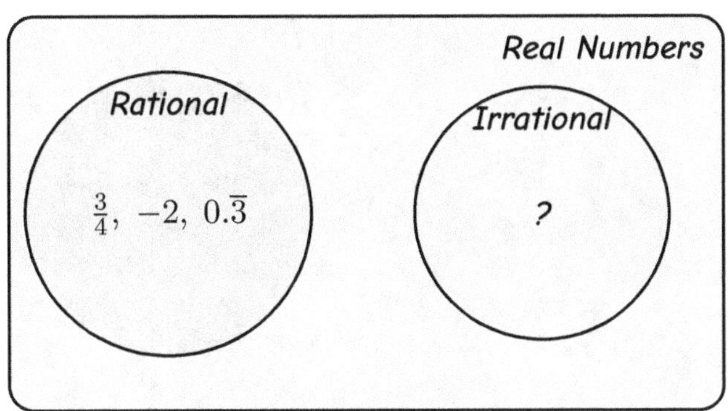

(A) Rational region, because $\sqrt{5} = 2.5$

(B) Irrational region, because 5 is not a perfect square

(C) Rational region, because $\sqrt{5}$ is a square root

(D) Neither region, because $\sqrt{5}$ is not a real number

2. The steps below show a student's work for converting $0.\overline{2}$ to a fraction. Which step contains the error?

Step	Work
1	Let $x = 0.222\ldots$
2	$10x = 2.222\ldots$
3	$10x - x = 2.222\ldots - 0.222\ldots$
4	$9x = 2$
5	$x = \frac{2}{10}$

(A) Step 2

(B) Step 3

(C) Step 4

(D) Step 5

3. Which is the best estimate for π^2?

(A) 6.3

(B) 9.9

(C) 3.1

(D) 12.6

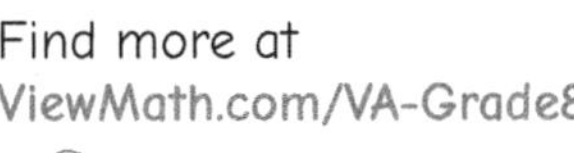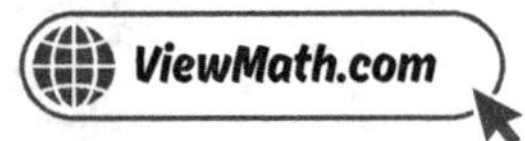

4. Which is a better deal: a $60 item with a 20% discount, or the same item at a different store for $50 with no discount?

(A) The 20% discount, because you pay $48.

(B) The $50 price, because $50 < $60.

(C) They cost the same.

(D) The 20% discount, because you pay $40.

5. Simplify $\frac{4^2 \cdot 4^{-5}}{4^{-1}}$.

(A) 4^{-2}

(B) 4^{-4}

(C) 4^6

(D) 4^{-8}

6. The diagram below shows three cells viewed under a microscope, with their actual sizes labeled. Write all three sizes in scientific notation, then list them from smallest to largest.

Cell A
0.008 mm

Cell B
0.05 mm

Cell C
0.0003 mm

Your Answer:

7. Compute $\frac{3.6 \times 10^{10}}{1.2 \times 10^4}$. Write your answer in scientific notation.

Your Answer:

8. If $y = kx$ and $y = 54$ when $x = 6$, find y when $x = 11$.

Your Answer:

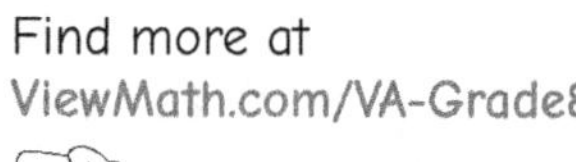
Find more at
ViewMath.com/VA-Grade8

ViewMath.com

9. How many solutions does $6x + 4 = 6x + 4$ have?

(A) No solution

(B) One solution

(C) Two solutions

(D) Infinitely many solutions

10. Pens cost \$2 and notebooks cost \$5. You buy 8 items for \$25. How many notebooks did you buy?

(A) 2

(B) 3

(C) 4

(D) 5

11. The graph below shows Function F. Function G is $y = 3x + 2$. Find the rate of change of each function and state which is greater.

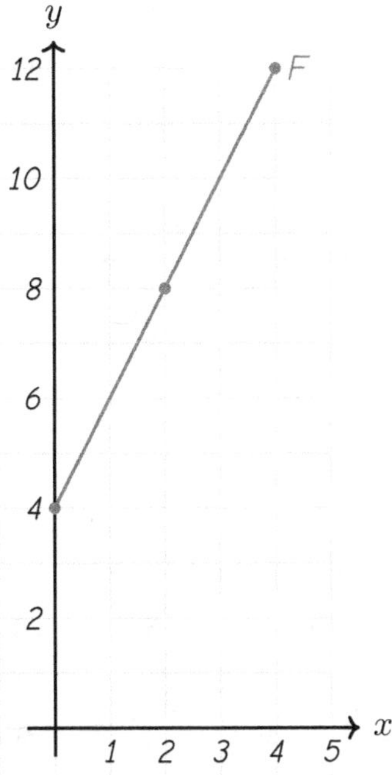

Your Answer:

Find more at
ViewMath.com/VA-Grade8

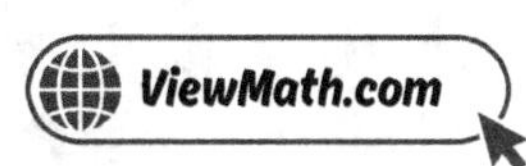

12. A table shows $(1, 5)$, $(2, 10)$, $(3, 15)$, $(4, 20)$. What is the rate of change, and is the function linear?

(A) Rate of change $= 5$; linear

(B) Rate of change $= 5$; nonlinear

(C) Rate of change varies; linear

(D) Rate of change varies; nonlinear

13. Use the graph below to write the equation of the line.

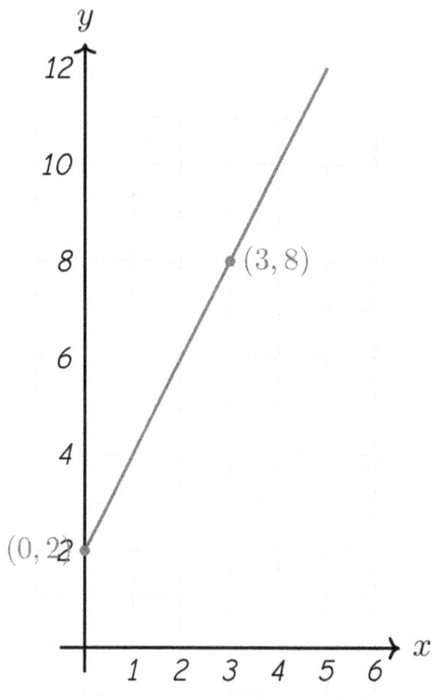

Your Answer:

14. Two distance-time graphs start at the same point. Graph A is steeper than Graph B. What can you say?

(A) Object A is farther from the start.

(B) Object A is traveling faster than Object B.

(C) Object B is traveling faster than Object A.

(D) Both objects travel at the same speed.

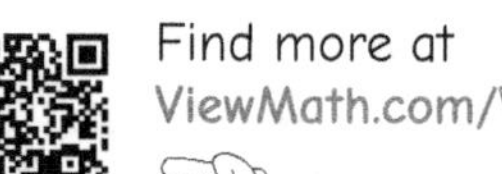

Find more at
ViewMath.com/VA-Grade8

ViewMath.com

15. A parallelogram is reflected over the y-axis. Which of the following changes?

(A) The side lengths

(B) The angle measures

(C) The orientation (left-right order of vertices)

(D) The area

16. $\triangle ABC \cong \triangle XYZ$. If $BC = 12$ cm, what is YZ?

Your Answer:

17. A rectangle has vertex $(-6, 2)$. After a translation of $(4, -5)$, what is the new position of this vertex?

(A) $(-10, 7)$

(B) $(-2, -3)$

(C) $(2, -3)$

(D) $(-2, 3)$

18. Triangle A has sides 3, 4, 5 as shown. Triangle B is similar to A with a scale factor of 2. What is the hypotenuse of Triangle B?

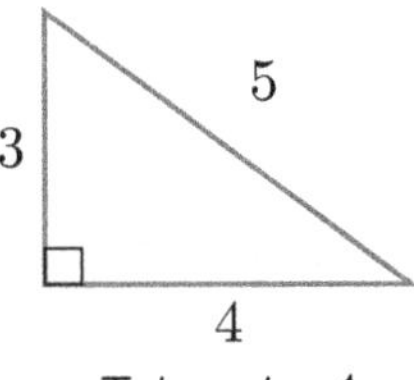

Your Answer:

19. Two parallel lines are cut by a transversal. Angle $1 = (3x + 10)°$ and Angle $2 = (5x - 30)°$ are alternate interior angles. What is x?

(A) 10

(B) 20

(C) 25

(D) 40

Find more at
ViewMath.com/VA-Grade8

ViewMath.com

20. *A right triangle has legs 9 and 12. What is the hypotenuse?*

(A) 21

(B) 108

(C) 15

(D) $\sqrt{21}$

21. *What is the volume of a cylinder with radius 4 cm and height 10 cm? Use $\pi \approx 3.14$.*

(A) 125.6 cm^3

(B) 502.4 cm^3

(C) 160 cm^3

(D) 251.2 cm^3

22. *Two lines intersect. One of the four angles is 42°. What are the other three angles?*

(A) 42°, 138°, 138°

(B) 42°, 42°, 42°

(C) 138°, 138°, 138°

(D) 48°, 42°, 48°

23. *A cube has edge 9 m. What is its surface area?*

Your Answer:

Find more at
ViewMath.com/VA-Grade8

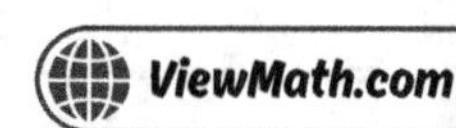

24. A square pyramid and a rectangular prism share the same base (6×6 in) and height (10 in). What is the volume of the pyramid?

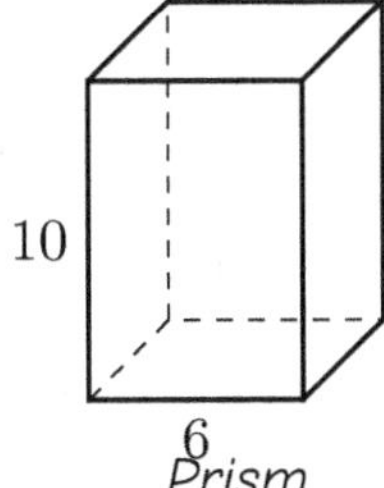 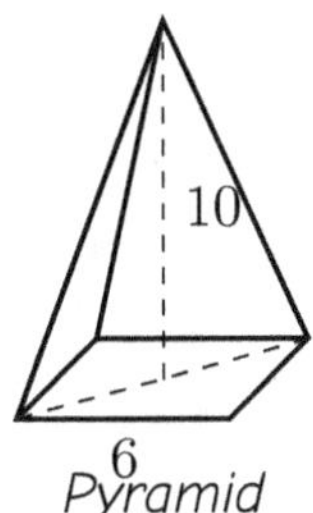

(A) 360 in^3

(B) 180 in^3

(C) 120 in^3

(D) 60 in^3

25. Two congruent rectangles, each 5×3 cm, overlap in a 3×3 square. What is the area of the combined figure?

(A) 30 cm^2

(B) 21 cm^2

(C) 24 cm^2

(D) 15 cm^2

26. A scatter plot of age (x) versus height (y) for people aged 0–80 would likely show what shape?

Your Answer:

27. When drawing an informal line of best fit, about how many data points should be above the line?

(A) All of them

(B) None of them

(C) About half

(D) Exactly one

28. A scatter plot with trend line $y = 2x + 1$ is shown. What is the predicted value at $x = 4$?

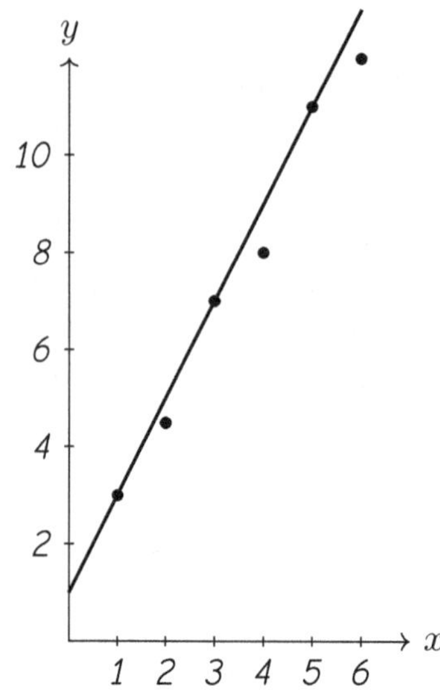

(A) 7

(B) 8

(C) 9

(D) 10

29.

	Pizza	Tacos	Total
Boys	24	16	40
Girls	18	22	40
Total	42	38	80

What is the relative frequency of boys who prefer pizza out of all 80 students?

(A) 0.30

(B) 0.60

(C) 0.20

(D) 0.525

30. A bag has 5 red and 3 blue marbles. One marble is drawn and replaced. Then another is drawn. What is $P(red, then\ blue)$?

(A) $\frac{15}{64}$

(B) $\frac{15}{56}$

(C) $\frac{5}{8}$

(D) $\frac{3}{8}$

Find more at
ViewMath.com/VA-Grade8

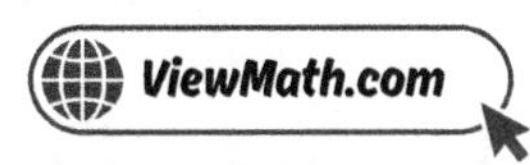

Great job finishing the test!

 My Score

I got _____________ out of 30 questions right.

Check your answers in the Answer Key at the back of the book.

💡 *Review any questions you missed. That's how we learn!*

📊 **Check Your Score Online!**

*Visit **ViewMath Academy** to enter your answers and see which topics you need to review. You can also explore lessons, take quizzes, track your scores, and save your progress!*

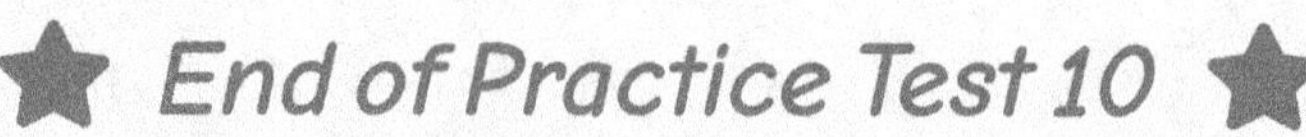

viewmath.com/score/8.1.VA.25

Or go to viewmath.com/score and enter code: 8.1.VA.25

Answer Key & Explanations

Answer Key

First try each test on your own, then check your work here.

✅ Practice Test 1 — Answer Key

 False    C    B    B    B   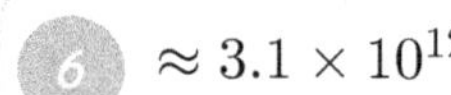 $\approx 3.1 \times 10^{12}$  C    B

 $x = 3$  80    B   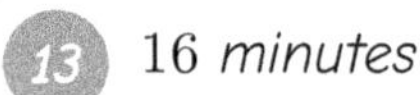 Nonlinear    16 minutes   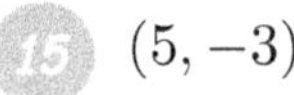 C    $(5, -3)$  C

17   $(-2, 5)$

18   1,500 m  B    C    C    B    B    C   25 B   26 B   27 B

28   At $x = 5$: $y = 30$. At $x = 30$: $y = -45$. The prediction at $x = 5$ is more reliable because it is within the data ran

29   C   30 $\frac{4}{52} \times \frac{3}{51} = \frac{12}{2652} = \frac{1}{221}$

💡 Time to Learn! 💡

Review the explanations below, **especially for the questions you missed**.

Understanding why each answer is correct builds stronger problem-solving skills.

Tip: Circle any questions you got wrong, then read their explanation carefully.

📖 Practice Test 1 — Detailed Explanations

Find more at
ViewMath.com/VA-Grade8

1 $\frac{22}{7} \approx 3.142857\ldots$ is a rational approximation of π, but $\pi = 3.14159265\ldots$ is irrational. They are close but not equal.

2 $0.\overline{123}$ has a 3-digit repeating block, so you should multiply by $10^3 = 1000$, not 100.

3 Area $= \sqrt{20} \times \sqrt{8} = \sqrt{160}$. $\sqrt{160} \approx 12.65$ cm^2. (Note: $\sqrt{160} = 4\sqrt{10} \approx 12.65$.) Choices A and C confuse addition with multiplication.

4 In $I = Prt$, P is the principal, r is the annual interest rate (as a decimal), and t is the time in years.

5 The star is at $\frac{1}{8}$. Since $\frac{1}{8} = \frac{1}{2^3} = 2^{-3}$, the answer is 2^{-3}.

6 $498{,}000{,}000 \approx 5 \times 10^8$ and $6{,}200 \approx 6.2 \times 10^3$. Product $\approx 5 \times 6.2 \times 10^{11} = 31 \times 10^{11} = 3.1 \times 10^{12}$.

7 $(3 \times 10^4)(5 \times 10^3) = 15 \times 10^7 = 1.5 \times 10^8$.

8 At \$8 per hour with no extra fee, the relationship is proportional: $y = 8x$.

9 Add $5x$: $14 = 4x + 2$. Subtract 2: $12 = 4x$. Divide by 4: $x = 3$.

10 $a + s = 200$ and $8a + 5s = 1240$. From first: $s = 200 - a$. Substitute: $8a + 5(200 - a) = 1240$, so $3a + 1000 = 1240$, $3a = 240$, $a = 80$.

11 Line A: slope $= \frac{7-1}{3-0} = 2$, initial value $= 1$. Line B: slope $= \frac{7-4}{3-0} = 1$, initial value $= 4$. A is steeper but starts lower.

12 Check the y-differences: $2 - 1 = 1$, $4 - 2 = 2$, $5 - 4 = 1$, $7 - 5 = 2$. The differences are not all the same $(1, 2, 1, 2)$, so the rate of change is not constant. The function is nonlinear.

13 $y = -5x + 80$. Set $y = 0$: $0 = -5x + 80 \Rightarrow 5x = 80 \Rightarrow x = 16$.

Find more at
ViewMath.com/VA-Grade8

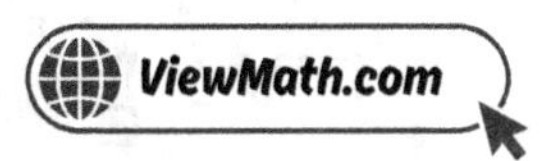

14 An upward graph means the output (distance) is getting larger as the input (time) increases — the object is moving away.

15 The 90° CCW rule is $(x, y) \rightarrow (-y, x)$. So $(-3, -5) \rightarrow (5, -3)$.

16 Two rectangles can have the same perimeter but different dimensions (e.g., 2×6 and 3×5 both have perimeter 16), so they are not necessarily congruent.

17 Multiply each coordinate by $\frac{1}{2}$: $(-4 \times \frac{1}{2}, \ 10 \times \frac{1}{2}) = (-2, 5)$.

18 Real length $= 3 \times 50,000 = 150,000 \ cm = 1,500 \ m$.

19 Alternate interior angles are equal: $4x = 80$, so $x = 20$.

20 $9^2 + 40^2 = 81 + 1600 = 1681 = 41^2$. The converse of the Pythagorean Theorem confirms it's a right triangle.

21 A cone's volume is $\frac{1}{3}$ of a cylinder with the same base and height. It takes exactly 3 cones to fill the cylinder.

22 Let the angles be x and $4x$. $x + 4x = 90$, so $5x = 90$ and $x = 18$. The larger angle is $4(18) = 72°$.

23 $SA = 2\pi r^2 + 2\pi rh = 2(3.14)(9) + 2(3.14)(3)(8) = 56.52 + 150.72 = 207.24 \ m^2$.

24 Prism volume $= 96 \ cm^3$. Pyramid volume $= \frac{1}{3}(16)(6) = 32 \ cm^3$. $96/32 = 3$.

25 Top rectangle $= 12 \times 2 = 24$. Bottom rectangle $= 4 \times 8 = 32$. Total $= 56 \ cm^2$.

26 The dots go from upper-left to lower-right in a roughly straight pattern. This is a negative linear association.

27 Outliers should generally be ignored when drawing the line of best fit, as they don't represent the overall trend.

28 $y = -3(5) + 45 = 30$ (interpolation). $y = -3(30) + 45 = -45$ (extrapolation — unreliable and gives a negative value that may not make sense).

29 $\frac{25}{40} = 0.625 = 62.5\%$.

30 First ace: $\frac{4}{52}$. Second ace (without replacement): $\frac{3}{51}$. Multiply: $\frac{12}{2652} = \frac{1}{221}$.

✅ Practice Test 2 — Answer Key

1 C   **2** Multiply by 100; $100x = 54.\overline{54}$; $99x = 54$; $x = \frac{54}{99} = \frac{6}{11}$ **3** ≈ 0.02 **4** B   **5** C

6 B   **7** A   **8** 180 mL   **9** C   **10** 5   **11** B   **12** Nonlinear; each output is x^3.

13 $y = 12x + 50$; $m = 12$ is the hourly wage, $b = 50$ is the base pay.   **14** Constant   **15** B   **16** B

17 B   **18** A   **19** 75°   **20** B   **21** B   **22** 115°   **23** A   **24** 100 cm^3 **25** B

26 Positive linear association. Outlier at approximately $(1, 7)$ — far above the trend.   **27** C   **28** C

29 B   **30** C

💡 Time to Learn! 💡

Review the explanations below, **especially for the questions you missed.**

Understanding why each answer is correct builds stronger problem-solving skills.

Tip: Circle any questions you got wrong, then read their explanation carefully.

Find more at
ViewMath.com/VA-Grade8

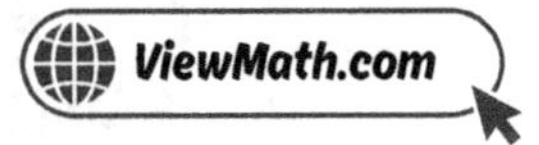

📖 Practice Test 2 — Detailed Explanations

1. $P = \sqrt{2}$ is irrational (since 2 is not a perfect square) and $Q = 3$ is an integer, which is rational.

2. The repeating block has 2 digits, so multiply by 100. Then $100x - x = 99x = 54$, giving $x = \frac{54}{99} = \frac{6}{11}$.

3. $\pi \approx 3.1416$ and $\sqrt{10} \approx 3.1623$. The distance is $3.1623 - 3.1416 = 0.0207 \approx 0.02$.

4. Markup $= 50 \times 0.40 = 20$. Selling price $= 50 + 20 = \$70$.

5. Numerator: $9^3 \cdot 9^2 = 9^5$. Then $\frac{9^5}{9^4} = 9^{5-4} = 9^1 = 9$.

6. $\frac{4.2 \times 10^3}{4.2 \times 10^{-3}} = 10^{3-(-3)} = 10^6$. So 4.2×10^3 is 10^6 (one million) times as large.

7. Divide the coefficients: $\frac{8}{2} = 4$. Subtract the exponents: $10^{7-3} = 10^4$. Answer: 4×10^4.

8. Rate $= \frac{45}{15} = 3$ mL/min. In 60 minutes: $3 \times 60 = 180$ mL.

9. Subtract 3: $\frac{x}{4} = 4$. Multiply by 4: $x = 16$.

10. $a + b = 12$ and $5a + 10b = 85$. From first: $a = 12 - b$. Substitute: $5(12 - b) + 10b = 85$, so $60 + 5b = 85$, $5b = 25$, $b = 5$.

11. $\frac{3}{4} = 0.75 > 0.5 = \frac{1}{2}$, so Function B is steeper.

12. $1^3 = 1$, $2^3 = 8$, $3^3 = 27$, $4^3 = 64$. The differences $(7, 19, 37)$ are not constant, confirming nonlinear.

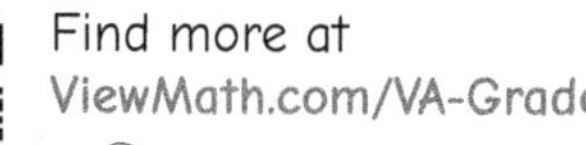

Find more at
ViewMath.com/VA-Grade8

13. *Slope:* $62 - 50 = 12$ *(hourly rate).* y-intercept: 50 *(base pay before any hours). Equation:* $y = 12x + 50$.

14. *A straight line has a constant slope, meaning the rate of change is the same everywhere along the line.*

15. *When reflecting over the* x*-axis, the* x*-coordinate stays the same and the* y*-coordinate flips sign:* $(3, 5) \to (3, -5)$.

16. *Equal angles make triangles similar, but not necessarily congruent — their sides could be different lengths.*

17. *A dilation by factor* k *from the origin gives* $(kx, ky) = (2 \times 3, 2 \times 4) = (6, 8)$.

18. *Perimeter scales by the same factor.* $36 \times \frac{1}{3} = 12$ *cm.*

19. $118 = 43 + x$, *so* $x = 118 - 43 = 75°$.

20. $7^2 + 24^2 = 49 + 576 = 625 = 25^2$. *Since* $a^2 + b^2 = c^2$, *it is a right triangle.*

21. $V = \frac{1}{3}(3.14)(64)(24) = \frac{1}{3}(4{,}823.04) \approx 1{,}607.7$ *in*^{3}.

22. $5x + 25 + 3x + 11 = 180$, *so* $8x + 36 = 180$, $8x = 144$, $x = 18$. *Larger:* $5(18) + 25 = 115°$.

23. *Base area* $= 10^2 = 100$. *Lateral SA* $= \frac{1}{2}P\ell = \frac{1}{2}(40)(13) = 260$. *Total* $= 100 + 260 = 360$ *ft*^{2}.

24. *Base area* $= \frac{1}{2}(5)(12) = 30$ *cm*^{2}. $V = \frac{1}{3}(30)(10) = 100$ *cm*^{3}.

25. *Split into bottom rectangle* $(12 \times 5 = 60)$ *and top-left rectangle* $(8 \times 3 = 24)$. *Total* $= 84$ *m*^{2}.

Find more at
ViewMath.com/VA-Grade8

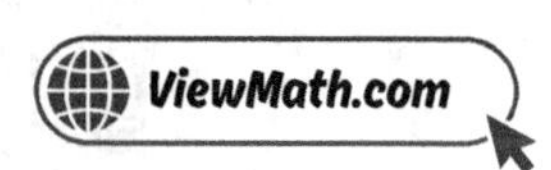

26 Direction: positive (up-right). Shape: linear. The point $(1, 7)$ is an outlier (most students who studied 1 hour scored around 3, not 7).

27 A line of best fit (trend line) is a straight line drawn to approximate the overall trend, coming close to most points.

28 A slope of 0 means no change in y as x changes; the line is horizontal.

29 60% of boys chose band vs. 45% of girls. The difference in conditional proportions shows an association.

30 2 coin outcomes $\times$ 6 die outcomes $= 12$ total outcomes.

✅ Practice Test 3 — Answer Key

 **1** C   **2** C   **3** C   **4** B   **5** B   **6** A   **7** A   **8** A   **9** B   **10** C

11 C   **12** Linear   **13** A   **14** Decreasing (fast at first, then slower), approaching constant.   **15**

16 C   **17** A   **18** B   **19** A   **20** B   **21** B   **22** $x = 20$ **23** B   **24** A   **25** B

26 Yes. Example: population growth over time may curve upward (exponential), showing a positive but nonlinear

27 B   **28** B   **29** B   **30** C

Find more at
ViewMath.com/VA-Grade8

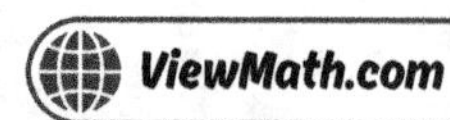

💡 Time to Learn! 💡

*Review the explanations below, **especially for the questions you missed.***

Understanding why each answer is correct builds stronger problem-solving skills.

***Tip:** Circle any questions you got wrong, then read their explanation carefully.*

📖 Practice Test 3 — Detailed Explanations

1. $0.\overline{81}$ is a repeating decimal, so it can be written as a fraction: $0.\overline{81} = \frac{81}{99} = \frac{9}{11}$. The others are irrational.

2. Let $x = 0.\overline{27}$. Then $100x = 27.\overline{27}$. Subtract: $99x = 27$, so $x = \frac{27}{99} = \frac{3}{11}$.

3. $4\sqrt{2} \approx 4 \times 1.414 = 5.656$ and $\sqrt{30} \approx 5.477$. Since $5.656 > 5.477$, $4\sqrt{2}$ is larger.

4. $I = 2000 \times 0.06 \times 2 = 240$. Total $= 2000 + 240 = \$2,240$.

5. When multiplying powers with the same base, you keep the base and add the exponents: $2^3 \cdot 2^4 = 2^7$, not 4^7. The student incorrectly multiplied the bases.

6. Move the decimal 4 places right: $0.00072 = 7.2 \times 10^{-4}$.

7. $(9.5 \times 10^{-18})(2 \times 10^6) = 19 \times 10^{-12} = 1.9 \times 10^{-11}$ g.

8. Store A: $k = 2$ per pound. Store B: $k = \frac{7.50}{3} = 2.50$ per pound. Store A is cheaper.

9. Subtract 1.5: $0.5x = 2.5$. Divide by 0.5: $x = 5$.

Find more at
ViewMath.com/VA-Grade8

10 $60t + 80t = 420$, so $140t = 420$, $t = 3$ hours.

11 $\frac{9-1}{2-0} = \frac{8}{2} = 4$. The rate of change is 4.

12 A constant increase of 6 means the rate of change is constant, which is the defining property of a linear function.

13 The slope $m = 50$ represents the amount added per week, so the weekly deposit is \$50.

14 The coffee cools quickly at first (steep decrease), then more slowly as it nears room temperature. The graph curves and levels off — nonlinear decreasing approaching constant.

15 A rotation turns (spins) a figure around a fixed point called the center of rotation.

16 Same shape but different size means the figures are similar (related by a dilation) but not congruent. Congruence requires equal size.

17 Check translation: $(-3 + 9,\ 4 + 4) = (6, 8)$. This matches a translation of $(9, 4)$.

18 When $k = 1$, every point stays in the same place. The image is congruent (identical in size and shape) to the original.

19 The exterior angle adjacent to the third angle equals the sum of the two non-adjacent interior angles: $37 + 53 = 90°$.

20 $a^2 = 17^2 - 8^2 = 289 - 64 = 225$, so $a = 15$. (Note: $\sqrt{225} = 15$, so B and D are the same value, but B is simplified.)

21 $200\pi = \pi r^2(8)$, so $r^2 = \frac{200}{8} = 25$, and $r = 5$ cm.

Find more at
ViewMath.com/VA-Grade8

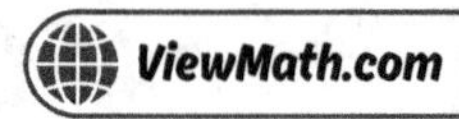

22 Vertical angles are equal: $3x + 15 = 5x - 25$, so $40 = 2x$ and $x = 20$.

23 Base area $= 12^2 = 144$. Lateral SA $= 360 - 144 = 216$. $\frac{1}{2}(48)\ell = 216$, so $24\ell = 216$ and $\ell = 9$ in.

24 $V = \frac{1}{3}(49)(12) = \frac{588}{3} = 196 \ cm^3$.

25 The key strategy for composite figures is to decompose them into familiar shapes (rectangles, triangles, circles, etc.) and then add or subtract their areas.

26 An association can be positive (both increase) and nonlinear (the trend is curved, not straight).

27 A straight line of best fit only works for data with a linear pattern. Curved data needs a different model.

28 $y = -4(10) + 100 = -40 + 100 = 60$.

29 The intersection of Girls row and Dog column is 12.

30 $P(6) = \frac{1}{6}$. Independent: $P = \frac{1}{6} \times \frac{1}{6} = \frac{1}{36}$.

✓ Practice Test 4 — Answer Key

 D    $\frac{53}{90}$ 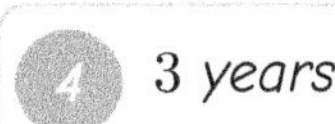 B    3 years    A    C    C    Taxi B    C

 C    B    4    C    Increasing, decreasing, increasing.    C    C

 B    C

 B    B    A    C    A    70 cm^3  B    C   27 C   28 B

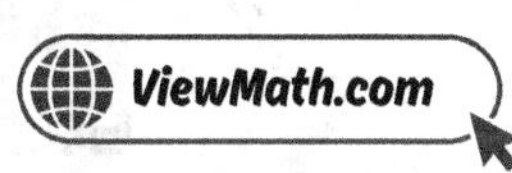

 **29** Yes. Of instrument players, $\frac{15}{40} = 37.5\%$ play a sport. Of non-instrument players, $\frac{30}{40} = 75\%$ play a sport. The la

 **30** A

💡 Time to Learn! 💡

Review the explanations below, **especially for the questions you missed.**

Understanding why each answer is correct builds stronger problem-solving skills.

Tip: Circle any questions you got wrong, then read their explanation carefully.

📖 Practice Test 4 — Detailed Explanations

1 $\sqrt{7}$ is irrational because 7 is not a perfect square. The other choices are all rational: $\frac{5}{8} = 0.625$, $0.\overline{6} = \frac{2}{3}$, and $\sqrt{9} = 3$.

2 Let $x = 0.5888\ldots$. Then $10x = 5.888\ldots$ and $100x = 58.888\ldots$. Subtract: $90x = 53$, so $x = \frac{53}{90}$.

3 $2\sqrt{5} = \sqrt{4} \cdot \sqrt{5} = \sqrt{4 \times 5} = \sqrt{20}$. Alternatively, $2\sqrt{5} \approx 4.47$ while $\sqrt{10} \approx 3.16$. They are not equal.

4 $I = Prt$, so $t = \frac{I}{Pr} = \frac{60}{400 \times 0.05} = \frac{60}{20} = 3$ years.

5 Using the product rule: $3^4 \cdot 3^2 = 3^{4+2} = 3^6$.

6 Move the decimal 3 places right: $6.1 \times 10^3 = 6,100$.

7 $\frac{2 \times 10^{-6}}{10^{-6}} = 2$. The bacterium is 2 micrometers long.

Find more at
ViewMath.com/VA-Grade8

8 Taxi A: \$3/mile. Taxi B: $\frac{36}{9} = \$4$/mile. Taxi B is more expensive per mile.

9 Multiply by 3: $2x - 1 = 15$. Add 1: $2x = 16$. Divide by 2: $x = 8$.

10 $x + y = 25$ and $x - y = 7$. Add: $2x = 32$, $x = 16$. The larger number is 16.

11 $M(0) = 7(0) + 2 = 2$. $N(0) = 10$ (given). Since $10 > 2$, Function N is greater at $x = 0$.

12 $6 - 2 = 4$, $10 - 6 = 4$, $14 - 10 = 4$. The rate of change is 4.

13 Slope: $10 - 7 = 3$. y-intercept: 7 (when $x = 0$). Equation: $y = 3x + 7$.

14 Filling: the water level rises. Drinking: the level drops. Refilling: the level rises again.

15 Reflecting over the y-axis flips the sign of the x-coordinate while keeping y the same: $(7, -2) \rightarrow (-7, -2)$.

16 The triangles have the same angles, making them similar. But the base of Triangle A is 6 cm while Triangle B's base is 3 cm, so they are not congruent.

17 $180°$ rotation: $(a, b) \rightarrow (-a, -b)$. Reflect over x-axis: $(-a, -b) \rightarrow (-a, b)$.

18 Multiply each dimension by 2.5: $4 \times 2.5 = 10$ in. $6 \times 2.5 = 15$ in.

19 Corresponding angles are in the same position (e.g., both upper-left) at each intersection of the transversal with the parallel lines.

20 $h^2 = 15^2 - 9^2 = 225 - 81 = 144$, so $h = 12$ ft.

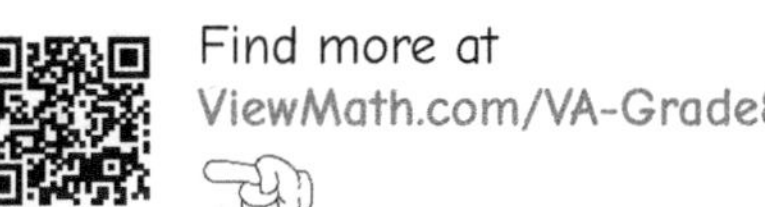
Find more at
ViewMath.com/VA-Grade8

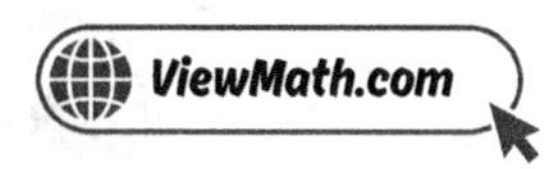

21 A cone is exactly $\frac{1}{3}$ the volume of a cylinder with the same base and height: $\frac{450\pi}{3} = 150\pi$ cm^3.

22 Vertical angles (across from each other when two lines cross) are always equal.

23 $SA = 2\pi(9) + 2\pi(3)(12) = 56.52 + 226.08 = 282.6$ cm^2.

24 $V = \frac{1}{3}(35)(6) = \frac{210}{3} = 70$ cm^3.

25 $12 \times 8 = 96.$ $Cut = 3^2 = 9.$ $Remaining = 96 - 9 = 87$ cm^2.

26 A curved pattern is a nonlinear association. The data has a trend, but it is not a straight line.

27 $y = 1.5(4) + 2 = 6 + 2 = 8.$

28 $x = 50$ is far outside the data range (1 to 10), so this is extrapolation.

29 Students who do NOT play an instrument are much more likely to play a sport (75% vs. 37.5%). This difference in conditional percentages signals an association.

30 With replacement: $\frac{6}{10} \times \frac{6}{10} = \frac{36}{100} = \frac{9}{25}.$

✅ Practice Test 5 — Answer Key

 Rational    B    A    B    B    11 places to the left    A    D

 B    Length = 16 m, width = 8 m    5    D   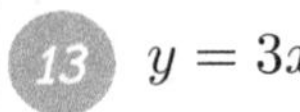 $y = 3x$  B    B

 55°    B    B    B    B    300 cm^3  B    B    B

Find more at
ViewMath.com/VA-Grade8

 A    C    Slope $= -1$; y-intercept $= 12$  A

 Totals: Grade A $= 30$, Not A $= 30$, HW Yes $= 40$, HW No $= 20$, Grand Total $= 60$. Relative frequencies: $\frac{24}{60} = 0.40$, $\frac{6}{60} = 0.$

B

💡 **Time to Learn!** 💡

Review the explanations below, **especially for the questions you missed.**

Understanding why each answer is correct builds stronger problem-solving skills.

Tip: Circle any questions you got wrong, then read their explanation carefully.

📖 **Practice Test 5 — Detailed Explanations**

1. $\sqrt{144} = 12$ exactly, because $12^2 = 144$. Since 12 is an integer, $\sqrt{144}$ is rational.

2. The repeating block has 2 digits, so multiply by $10^2 = 100$. Then $100x = 63.6363\ldots$ and $x = 0.6363\ldots$. Subtracting: $100x - x = 63$.

3. Perimeter $= 4 \times \sqrt{50} = 4\sqrt{50} \approx 4 \times 7.07 = 28.28 \approx 28.3$ cm. Note that $\sqrt{200} \approx 14.14$, which is only $2\sqrt{50}$ — that would be two sides, not four.

4. Tax $= 65 \times 0.08 = 5.20$. Total $= 65 + 5.20 = \$70.20$.

5. Using the power rule: $(2^3)^4 = 2^{3 \times 4} = 2^{12}$.

6. The exponent is -11, so you move the decimal 11 places to the left from 2.5.

Find more at
ViewMath.com/VA-Grade8

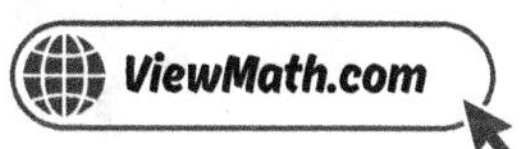

7. Same exponent, so add the coefficients: $5.2 + 3.8 = 9.0$. Answer: 9×10^6.

8. If the relationship passes through $(1,5)$ and $(3,15)$, then $k = 5$. But $5 \times 2 = 10 \neq 13$, so $(2,13)$ does not fit.

9. Subtract 8: $-3x = -6$. Divide by -3: $x = 2$.

10. $2l + 2w = 48$ and $l = 2w$. Substitute: $2(2w) + 2w = 48$, $6w = 48$, $w = 8$. Then $l = 16$.

11. $\frac{19-4}{3-0} = \frac{15}{3} = 5$.

12. $y = x^2 - 1$ has x raised to the 2nd power, making it nonlinear. The others are all in the form $y = mx + b$.

13. Slope: $\frac{12-3}{4-1} = \frac{9}{3} = 3$. Using $(1,3)$: $3 = 3(1) + b \Rightarrow b = 0$. Equation: $y = 3x$.

14. Accelerating: speed increases. Cruising: speed stays the same (constant). Braking: speed decreases.

15. The $90°$ CCW rule is $(x,y) \to (-y,x)$. So $(2,0) \to (0,2)$.

16. $\angle R = 180 - 72 - 53 = 55°$. Since $\angle R$ corresponds to $\angle U$, $\angle U = 55°$.

17. Divide the image coordinates by the scale factor: $(9 \div 3,\ -15 \div 3) = (3,-5)$.

18. When $k > 1$, the dilation enlarges the figure. Each side length is multiplied by 3, making the figure 3 times larger in each dimension.

19. $180 - 54 - 54 = 72°$.

20. $c^2 = 1^2 + 1^2 = 2$, so $c = \sqrt{2}$.

Find more at
ViewMath.com/VA-Grade8

ViewMath.com

21 A cylinder is 3 times the volume of a cone with the same base and height: $100 \times 3 = 300 \ cm^3$.

22 Angles on a straight line are supplementary: $3x + (x + 20) = 180$, so $4x + 20 = 180$, $4x = 160$, $x = 40$.

23 Each face area appears twice: $SA = 2(15 + 20 + 12) = 2(47) = 94 \ cm^2$.

24 $V = \frac{1}{3}Bh = \frac{1}{3}(36)(10) = 120 \ cm^3$.

25 $Circle = \pi(36) = 113.04$. $Square = 36$. $Remaining = 113.04 - 36 = 77.04 \ cm^2$.

26 The independent (explanatory) variable goes on the x-axis; the dependent (response) variable goes on the y-axis.

27 $m = \frac{4-10}{8-2} = \frac{-6}{6} = -1$. $10 = -1(2) + b$, so $b = 12$.

28 $35 = 4x + 7$, $4x = 28$, $x = 7$.

29 Add rows and columns for totals. Divide each cell by 60 for relative frequency.

30 $P(A \text{ and } B) = 0.6 \times 0.4 = 0.24$.

✅ Practice Test 6 — Answer Key

 Rational  C  D  B  B  B  A  B  C

 C  B  B  7  B  B  C  (1, 2) 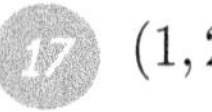 False

 B  $a^2 + 16a^2 = 17a^2$ 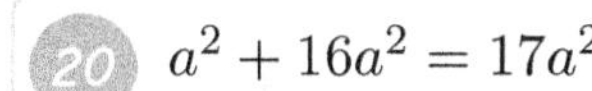 B  10  C  B  A  B  A

Find more at
ViewMath.com/VA-Grade8

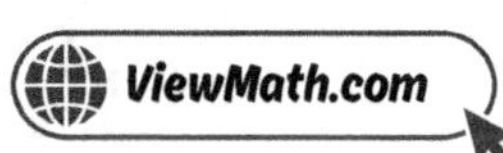

 **28** B    **29** B    **30** C

> 💡 **Time to Learn!** 💡
>
> *Review the explanations below, **especially for the questions you missed**.*
>
> *Understanding why each answer is correct builds stronger problem-solving skills.*
>
> ***Tip:*** *Circle any questions you got wrong, then read their explanation carefully.*

📖 Practice Test 6 — Detailed Explanations

1. $\frac{5}{6}$ is a fraction of two integers with a nonzero denominator, so it is rational. Its decimal is $0.8\overline{3}$, which repeats.

2. The repeating block 123 has 3 digits, so you multiply by $10^3 = 1000$.

3. $\pi \approx 3.1416$, $\sqrt{10} \approx 3.1623$, and 3.2 is exact. From least to greatest: $3.1416 < 3.1623 < 3.2$.

4. $I = 1000 \times 0.03 \times 4 = 120$. Total $= 1000 + 120 = \$1{,}120$.

5. $\frac{8^6}{8^6} = 8^{6-6} = 8^0 = 1$. The expression equals 8^0.

6. Move the decimal 5 places left: $9.03 \times 10^{-5} = 0.0000903$.

7. $4 \times (9.5 \times 10^{12}) = 38 \times 10^{12} = 3.8 \times 10^{13}$ km.

8. $k = \frac{y}{x} = \frac{9}{2} = 4.5$.

Find more at
ViewMath.com/VA-Grade8

9 $2x + 10 = 3x - 1$. Subtract $2x$: $10 = x - 1$. Add 1: $x = 11$.

10 $n + d = 15$ and $5n + 10d = 110$. From first: $n = 15 - d$. Substitute: $5(15 - d) + 10d = 110$, so $75 + 5d = 110$, $5d = 35$, $d = 7$.

11 $G(4) = 14$. $H(4) = 4(4) + 1 = 17$. Since $17 > 14$, Function H has the greater value at $x = 4$.

12 Graph A is a curve (parabola) — nonlinear. Graph B is a straight line — linear.

13 $1 = -3(2) + b \Rightarrow 1 = -6 + b \Rightarrow b = 7$.

14 Losing fuel at a steady rate means the amount decreases at a constant rate — that's a straight line going down (linear and decreasing).

15 Translations preserve all side lengths. The longest side is still 8 cm.

16 Two circles with the same radius are always congruent — one can be translated to overlap the other.

17 Add the translation: $(4 + (-3), \ -6 + 8) = (1, 2)$.

18 A 2×5 rectangle and a 3×4 rectangle have the same angles ($90°$) but their side ratios differ ($\frac{2}{5} \neq \frac{3}{4}$), so they are not similar.

19 $x + 2x + 3x = 180$, so $6x = 180$ and $x = 30$.

20 $a^2 + (4a)^2 = a^2 + 16a^2 = 17a^2 = (\sqrt{17} \cdot a)^2$. Confirmed.

21 $V = \frac{1}{3}\pi r^2 h = \frac{1}{3}\pi(9)(12) = \frac{108\pi}{3} = 36\pi \ m^3$.

Find more at
ViewMath.com/VA-Grade8

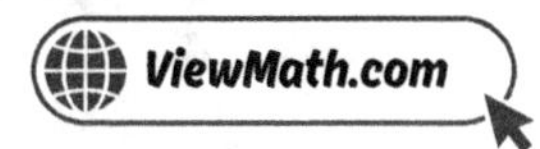

22. $4x + 8 = 6x - 12$, so $20 = 2x$ and $x = 10$.

23. $SA = 4 \times 25 = 100\ cm^2$.

24. $V = \frac{1}{3}Bh = \frac{1}{3}(30)(9) = 90\ cm^3$.

25. Rectangle $= 40$. Semicircle $= \frac{1}{2}\pi(4)^2 = \frac{1}{2}(50.24) = 25.12$. Total $= 65.12\ cm^2$.

26. As x increases, y increases. The dots go up-right, indicating a positive association.

27. $m = \frac{8-3}{5-1} = \frac{5}{4}$.

28. Extrapolation means using the model to predict values outside the data range, which can be unreliable.

29. The 20 is the count for a specific combination of two variables (7th grade AND soccer), making it a joint frequency.

30. An impossible event has probability 0 because it can never occur.

✅ Practice Test 7 — Answer Key

 $\sqrt{2}$ (or $\sqrt{3}$)  C  ≈ 4.4  B  7^2  B  317 times  C

 B  C  Yes  B 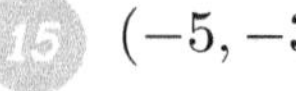 C  C  $(-5, -3)$  B  B

 D  B  26  B  B  $100\ cm^2$  B  $80\ cm^2$  C

 $y = -1.5x + 9.5$ (approximately)  B  B  C

Find more at
ViewMath.com/VA-Grade8

💡 Time to Learn! 💡

*Review the explanations below, **especially for the questions you missed**.*

Understanding why each answer is correct builds stronger problem-solving skills.

***Tip:** Circle any questions you got wrong, then read their explanation carefully.*

📖 Practice Test 7 — Detailed Explanations

1. $\sqrt{2} \approx 1.414$ is between 1 and 2 and is irrational because 2 is not a perfect square. $\sqrt{3} \approx 1.732$ also works.

2. $\frac{36}{100} = 0.36$ (terminates). The correct conversion uses $99x = 36$, giving $\frac{36}{99} = \frac{4}{11}$. The student treated $0.\overline{36}$ as if it were 0.36.

3. $\sqrt{3} \approx 1.732$ and $\sqrt{7} \approx 2.646$. Adding: $1.732 + 2.646 = 4.378 \approx 4.4$.

4. $I = Prt$, so $r = \frac{I}{Pt} = \frac{96}{800 \times 4} = \frac{96}{3200} = 0.03 = 3\%$.

5. $\frac{7^3 \cdot 7^4}{7^5} = \frac{7^7}{7^5} = 7^{7-5} = 7^2$.

6. Move the decimal 5 places left: $350{,}000 = 3.5 \times 10^5$. In scientific notation, a must satisfy $1 \le a < 10$.

7. $\frac{1.9 \times 10^{27}}{6 \times 10^{24}} = \frac{1.9}{6} \times 10^3 \approx 0.317 \times 10^3 = 317$.

8. In choice C, $\frac{9}{3} = 3$ but $\frac{15}{6} = 2.5$. The ratio is not constant, so it is not proportional.

9. Subtract 7: $3x = 15$. Divide by 3: $x = 5$.

Find more at
ViewMath.com/VA-Grade8

10 $2l + 2w = 34$ and $l = w + 5$. Substitute: $2(w + 5) + 2w = 34$, so $4w + 10 = 34$, $4w = 24$, $w = 6$.

11 Lines with the same slope but different y-intercepts are parallel.

12 For a function to be linear, x must have an exponent of 1. Since x is squared in $y = 3x^2$, the function is nonlinear regardless of the coefficient.

13 $m = -5$ and the y-intercept is 30 (given by the point $(0, 30)$). So $y = -5x + 30$.

14 The four sections are: increasing, decreasing, increasing, constant.

15 Reflecting over the y-axis flips the sign of the x-coordinate: $(5, -3) \rightarrow (-5, -3)$.

16 All corresponding sides are equal. Since the side lengths match, a rigid transformation can map one onto the other, so they are congruent.

17 Add the translation to each coordinate: $(5 + (-2),\ -3 + 4) = (3, 1)$.

18 Set up the proportion: $\frac{8}{x} = \frac{2}{5}$. Cross-multiply: $2x = 40$, so $x = 20$ cm.

19 One angle is $90°$ and another is $32°$. Third angle: $180 - 90 - 32 = 58°$.

20 $c^2 = 10^2 + 24^2 = 100 + 576 = 676$, so $c = \sqrt{676} = 26$.

21 $V = \pi r^2 h = \pi(4)(7) = 28\pi$ cm^3.

22 $(2x + 15) + (3x - 10) = 180$. Simplify: $5x + 5 = 180$, $5x = 175$, $x = 35$.

Find more at
ViewMath.com/VA-Grade8

ViewMath.com

23 $SA = 2(28) + 2(14) + 2(8) = 56 + 28 + 16 = 100 \ cm^2$.

24 $V = \frac{1}{3}(12 \times 5)(8) = \frac{1}{3}(60)(8) = \frac{480}{3} = 160 \ ft^3$.

25 Two rectangles $= 2(48) = 96$. Subtract overlap $= 16$. Area $= 96 - 16 = 80 \ cm^2$.

26 A strong positive trend corresponds to an r-value close to 1. $r = 0.9$ indicates a strong positive association.

27 Using $(1, 8)$ and $(5, 2)$: slope $= \frac{2-8}{5-1} = \frac{-6}{4} = -1.5$. $8 = -1.5(1) + b$, $b = 9.5$. Line: $y = -1.5x + 9.5$.

28 In $y = mx + b$, the slope m represents the rate of change: how much y changes for each 1-unit increase in x.

29 Of 42 pizza lovers, 24 are boys. $\frac{24}{42} \approx 0.571$.

30 There are 13 hearts in 52 cards: $P = \frac{13}{52} = \frac{1}{4}$.

✓ Practice Test 8 — Answer Key

1 C   **2** $\frac{81}{99} = \frac{9}{11}$ **3** ≈ 8.7 feet   **4** A   **5** 1.2   **6** C   **7** B   **8** A   **9** C

10 16 and 20   **11** B   **12** C   **13** 4

14 The runner jogs for 3 minutes (increasing), stops for 2 minutes (constant), jogs again for 3 minutes (increasing), then sto

15 Translation, reflection, rotation   **16** 30 cm   **17** $(-5, -2)$ **18** C   **19** 25   **20** C

21 $\approx 226.08 \ in^3$ **22** A   **23** A   **24** 6 in   **25** 27.42 cm   **26** C

27 Approximately 2 to 2.25

28 C

29 Boys: $\frac{20}{30} \approx 66.7\%$. Girls: $\frac{12}{20} = 60\%$. The percentages are close, so association is weak or none.

30 B

💡 Time to Learn! 💡

Review the explanations below, **especially for the questions you missed**.

Understanding why each answer is correct builds stronger problem-solving skills.

Tip: Circle any questions you got wrong, then read their explanation carefully.

📖 Practice Test 8 — Detailed Explanations

1 Every integer n can be written as $\frac{n}{1}$, making it rational. Not every square root is irrational (e.g., $\sqrt{4} = 2$).

2 Following the same pattern: $99x = 81$, so $x = \frac{81}{99}$. Simplify by dividing both by 9: $\frac{81}{99} = \frac{9}{11}$.

3 Side $= \sqrt{75}$. $8.6^2 = 73.96$ and $8.7^2 = 75.69$. Since 75 is between these, $\sqrt{75} \approx 8.7$.

4 $I = Prt$, so $P = \frac{I}{rt} = \frac{45}{0.03 \times 5} = \frac{45}{0.15} = 300$. The principal was $300.

5 $5^0 = 1$ and $5^{-1} = \frac{1}{5} = 0.2$. So $1 + 0.2 = 1.2$.

6 $1 \times 10^2 = 100$ kg. From the chart, the lion's bar reaches close to the 10^2 mark, making its mass closest to 100 kg.

7 $\frac{6.4 \times 10^{10}}{4 \times 10^6} = 1.6 \times 10^4 = 16{,}000$ photos.

Find more at
ViewMath.com/VA-Grade8

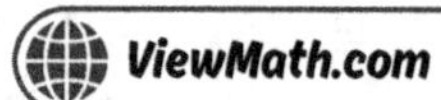
ViewMath.com

8   Machine A: 12 bottles/hour. Machine B: $\frac{50}{5} = 10$ bottles/hour. Machine A is faster.

9   A true statement like $5 = 5$ means every value of x satisfies the equation.

10 $x = y + 4$ and $x + y = 36$. Substitute: $(y + 4) + y = 36$, $2y = 32$, $y = 16$. Then $x = 20$.

11 $P(0) = 5$. For Function Q, $b = 9$. Since $9 > 5$, Function Q has the greater initial value.

12 $y = -2x + 6$ fits the form $y = mx + b$ with $m = -2$ and $b = 6$. A negative slope does not make a function nonlinear.

13 $\frac{25-9}{6-2} = \frac{16}{4} = 4$

14   0–3: distance increases steadily (jogging). 3–5: flat (resting). 5–8: distance increases again (jogging). 8–10: flat (stopped).

15   The three rigid transformations are translation (slide), reflection (flip), and rotation (turn). They all preserve size and shape.

16   Congruent figures have equal perimeters because all corresponding sides are equal.

17   180° rotation: $(x, y) \rightarrow (-x, -y)$. So $(5, 2) \rightarrow (-5, -2)$.

18   All squares have four 90° angles. Any two squares can be mapped by a dilation since the ratio of their sides is constant. So all squares are similar.

19 $(2x + 5) + 3x + (x + 25) = 180$. Combine: $6x + 30 = 180$, $6x = 150$, $x = 25$.

20 $6^2 + 8^2 = 36 + 64 = 100$, but $11^2 = 121$. Since $100 \neq 121$, this is not a right triangle.

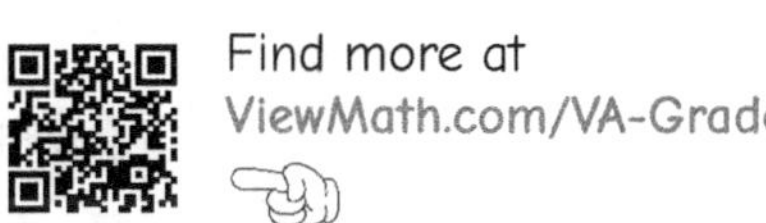

Find more at
ViewMath.com/VA-Grade8

21 $V = 3.14 \times 9 \times 8 = 226.08 \ in^3.$

22 Supplementary angles add to $180°$: $180 - 125 = 55°.$

23 Lateral $SA = 2\pi rh = 2(3.14)(7)(10) = 439.6 \ cm^2.$

24 $48 = \frac{1}{3}(s^2)(4),\ 48 = \frac{4s^2}{3},\ 144 = 4s^2,\ s^2 = 36,\ s = 6 \ in.$

25 Three straight sides $= 3(6) = 18.$ Semicircle circumference $= \frac{1}{2}\pi d = \frac{1}{2}(3.14)(6) = 9.42.$ Perimeter $= 18 + 9.42 = 27.42 \ cm.$

26 When data points group closely together, it is called a cluster.

27 Using endpoints: $m \approx \frac{11-2}{5-1} = \frac{9}{4} = 2.25.$ A slope around 2 is a reasonable estimate.

28 For each 1-unit increase in x, y increases by the slope (6). For 3 units: $6 \times 3 = 18.$

29 Compare the conditional relative frequencies. Since 66.7% and 60% are similar, the preference doesn't change much by gender.

30 $P(not\ event) = 1 - P(event) = 1 - \frac{3}{4} = \frac{1}{4}.$

☑ Practice Test 9 — Answer Key

 C  B  B  A  C  B  C 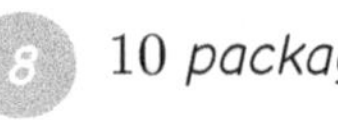 10 packages  B

 7 T-shirts and 8 hats  C  C  A  C  A  B  B

Find more at
ViewMath.com/VA-Grade8

 **C**    **A**    **B**    **C**    **B**    **A**    **B**    **B**    **A**    **A**

 **B**    **C**

30   Independent: flipping a coin twice (first flip doesn't affect second). Dependent: drawing cards without replacement (first

---

💡 Time to Learn! 💡

Review the explanations below, **especially for the questions you missed**.

Understanding why each answer is correct builds stronger problem-solving skills.

Tip: Circle any questions you got wrong, then read their explanation carefully.

📖 Practice Test 9 — Detailed Explanations

1 $1.41421356\ldots$ is the decimal expansion of $\sqrt{2}$, which is non-repeating and non-terminating, indicating an irrational number.

2   Let $x = 0.444\ldots$. Then $10x = 4.444\ldots$. Subtract: $9x = 4$, so $x = \frac{4}{9}$. Note that $\frac{4}{10} = 0.4$ (terminates), which is different from $0.\overline{4}$.

3 $\sqrt{2} \approx 1.414$, so $3\sqrt{2} \approx 4.243$. This is between 4 and 5.

4   Tip $= 48 \times 0.15 = 7.20$. The tip is $7.20.

5 $10^{-4} \cdot 10^{7} = 10^{-4+7} = 10^{3}$.

6 $8 \times 10^{5} = 800{,}000$ and $3 \times 10^{6} = 3{,}000{,}000$. The higher exponent wins: 3×10^{6} is larger.

Find more at
ViewMath.com/VA-Grade8

7 $7 \times 8 = 56$ and $10^{-3} \times 10^{-2} = 10^{-5}$. Then $56 \times 10^{-5} = 5.6 \times 10^{-4}$.

8 Service P: $4 \times 10 = 40$ packages. Service Q: $k = \frac{10}{2} = 5$, so $5 \times 10 = 50$ packages. Difference: $50 - 40 = 10$.

9 $4x - 12 = 2x + 2$. Subtract $2x$: $2x - 12 = 2$. Add 12: $2x = 14$, so $x = 7$.

10 $t + h = 15$ and $12t + 8h = 148$. From first: $h = 15 - t$. Substitute: $12t + 8(15 - t) = 148$, $4t + 120 = 148$, $4t = 28$, $t = 7$. Then $h = 8$.

11 Function R: when $x = 0$, $y = 1$. Function S: $b = 1$. Both start at 1, so the initial values are equal.

12 A linear function has the form $y = mx + b$. Only $y = 4x - 7$ fits this form with $m = 4$ and $b = -7$.

13 Slope: $\frac{16-7}{6-3} = \frac{9}{3} = 3$. Using $(3, 7)$: $7 = 3(3) + b \Rightarrow 7 = 9 + b \Rightarrow b = -2$.

14 A horizontal line means the output does not change. The height remains constant during that time period.

15 Add the translation to each coordinate: $(1 + 3, 2 + 1) = (4, 3)$.

16 When triangles are congruent, all corresponding sides are equal. Since AB corresponds to DE, $DE = 8$ cm.

17 Reflecting over the x-axis keeps the x-coordinate and negates the y-coordinate: $(x, y) \to (x, -y)$.

18 Check ratios: $\frac{6}{3} = 2$ but $\frac{9}{5} = 1.8$. Since $2 \neq 1.8$, the sides are not proportional and the rectangles are not similar.

19 An exterior angle and its adjacent interior angle are supplementary: $180 - 140 = 40°$.

20 $c^2 = 10^2 + 10^2 = 200$, so $c = \sqrt{200} = 10\sqrt{2}$.

21 $V = \frac{4}{3}\pi(3r)^3 = \frac{4}{3}\pi(27r^3) = 27 \times \frac{4}{3}\pi r^3$. The volume is 27 times as large.

22 Complementary angles add to $90°$: $90 - 37 = 53°$.

23 $SA = 2(12 \times 5) + 2(12 \times 3) + 2(5 \times 3) = 120 + 72 + 30 = 222 \ m^2$.

24 $V = \frac{1}{3}Bh = \frac{1}{3}(40)(12) = 160 \ m^3$.

25 Bottom rectangle $= 10 \times 6 = 60$. Top rectangle $= 4 \times 3 = 12$. Total $= 72 \ cm^2$.

26 Clusters are groups of points that bunch together, while outliers are isolated far from the trend. Both can appear in the same scatter plot.

27 The line goes from about $(0.5, 6.5)$ to $(7.5, 1.5)$. Slope $\approx \frac{1.5 - 6.5}{7.5 - 0.5} = \frac{-5}{7} \approx -0.71$.

28 $y = 5(8) + 20 = 40 + 20 = 60$ dollars.

29 Two-way tables deal with categorical data (counts and proportions). Mean scores involve numerical data, not categorical.

30 Independent events don't affect each other's probabilities. Dependent events do — the first outcome changes the conditions for the second.

✅ Practice Test 10 — Answer Key

 B　　 D　　3 B　　 A　　5 A　　6 Cell C, Cell A, Cell B　　7 3×10^6　　 99

 D　　 B　　 F: rate of change = 2; G: rate of change = 3. Function G is greater.　　 A

 $y = 2x + 2$　　 B　　15 C　　 12 cm　　 B　　 10　　19 B　　20 C　　21 B

22 A　　23 486 m^2　　24 C　　25 B

26 Nonlinear — height increases in childhood, then levels off in adulthood.　　27 C　　28 C　　29 A

30 A

💡 Time to Learn! 💡

*Review the explanations below, **especially for the questions you missed**.*

Understanding why each answer is correct builds stronger problem-solving skills.

Tip: *Circle any questions you got wrong, then read their explanation carefully.*

📖 Practice Test 10 — Detailed Explanations

1. 5 is not a perfect square, so $\sqrt{5}$ is irrational and belongs in the Irrational region.

2. In Step 5, the student divided 2 by 10 instead of 9. From $9x = 2$, the correct answer is $x = \frac{2}{9}$.

3. $\pi \approx 3.14$, so $\pi^2 \approx 3.14 \times 3.14 = 9.8596 \approx 9.9$.

4. $60 \times 0.20 = 12$ discount, so the price is $60 - 12 = \$48$. Since $\$48 < \$50$, the discount is the better deal.

5. Numerator: $4^2 \cdot 4^{-5} = 4^{-3}$. Then $\frac{4^{-3}}{4^{-1}} = 4^{-3-(-1)} = 4^{-2}$.

Find more at
ViewMath.com/VA-Grade8

6   Cell A: 8×10^{-3} mm. Cell B: 5×10^{-2} mm. Cell C: 3×10^{-4} mm. From smallest to largest: $3 \times 10^{-4} < 8 \times 10^{-3} < 5 \times 10^{-2}$, so Cell C, Cell A, Cell B.

7 $\frac{3.6}{1.2} = 3$ and $10^{10-4} = 10^6$. Answer: 3×10^6.

8 $k = \frac{54}{6} = 9$. When $x = 11$: $y = 9 \times 11 = 99$.

9   Subtracting $6x + 4$ from both sides gives $0 = 0$, which is always true. Every value of x is a solution.

10 $p + n = 8$ and $2p + 5n = 25$. From first: $p = 8 - n$. Substitute: $2(8 - n) + 5n = 25$, so $16 + 3n = 25$, $3n = 9$, $n = 3$.

11   From the graph, F rises from $(0, 4)$ to $(2, 8)$: slope $= \frac{8-4}{2-0} = 2$. Function G has slope 3. Since $3 > 2$, G has the greater rate of change.

12 $10 - 5 = 5$, $15 - 10 = 5$, $20 - 15 = 5$. The constant rate of change of 5 confirms the function is linear.

13 y-intercept is 2. Slope: $\frac{8-2}{3-0} = \frac{6}{3} = 2$. Equation: $y = 2x + 2$.

14   A steeper distance-time graph means more distance covered per unit of time — that means greater speed.

15   A reflection reverses the orientation (like looking at a mirror image). Side lengths, angles, and area are all preserved.

16   Corresponding sides of congruent triangles are equal. BC corresponds to YZ, so $YZ = 12$ cm.

17   Add the translation: $(-6 + 4, \ 2 + (-5)) = (-2, -3)$.

18   The hypotenuse of Triangle A is 5. Multiply by the scale factor: $5 \times 2 = 10$.

Find more at
ViewMath.com/VA-Grade8

19 Alternate interior angles are equal: $3x + 10 = 5x - 30$, so $40 = 2x$ and $x = 20$.

20 $c^2 = 9^2 + 12^2 = 81 + 144 = 225$, so $c = 15$.

21 $V = \pi r^2 h = 3.14 \times 16 \times 10 = 502.4 \; cm^3$.

22 Vertical angles: $42°$ across from $42°$. Adjacent supplementary angles: $180 - 42 = 138°$. So the four angles are $42°, 138°, 42°, 138°$.

23 $SA = 6(81) = 486 \; m^2$.

24 Prism volume $= 6 \times 6 \times 10 = 360 \; in^3$. Pyramid volume $= \frac{360}{3} = 120 \; in^3$.

25 Total without overlap $= 2(15) = 30$. Subtract overlap $= 9$. Combined $= 30 - 9 = 21 \; cm^2$.

26 Height rises quickly in youth, slows, and plateaus in adulthood. This is a nonlinear pattern.

27 A good line of best fit has roughly half the points above and half below.

28 $y = 2(4) + 1 = 9$.

29 $\frac{24}{80} = 0.30$.

30 With replacement: $P(red) = \frac{5}{8}$, $P(blue) = \frac{3}{8}$. $P = \frac{5}{8} \times \frac{3}{8} = \frac{15}{64}$.

Well done checking your answers!

Keep practicing to strengthen your skills.

Find more at
ViewMath.com/VA-Grade8

Author's Final Note

I hope you enjoyed this book as much as I enjoyed writing it. Whether you are a student working through the material, a parent supporting your child's learning, or a teacher guiding your class, I have tried to make this book as clear and engaging as possible. I hope I have succeeded. If you have any suggestions for improvement, please let me know. I would love to hear from you.

The accuracy of calculations is very important to me. We have done our best, but I also expect that I have made some minor errors. Constant improvement is the name of the game. If you find any errors, please let me know. I will fix them in the next edition.

For students: Your learning journey does not end here. I have written a series of books to help you learn math. Make sure you browse through them. I especially recommend workbooks and practice tests to help you prepare for your exams.

For parents: Thank you for investing in your child's education. I encourage you to explore the companion resources available online to help support your child outside the classroom.

For teachers: Thank you for the invaluable work you do every day. I hope this book serves as a useful resource in your classroom. Feel free to reach out if you have suggestions or would like to discuss how best to use this book with your students.

I also enjoy reading your reviews. If you have a moment, please leave a review on where you found this book. It will help others find this book. If you have any questions or comments, please feel free to contact me at Reza.Nazari@ViewMath.com.

And one last thing: Remember to use online resources for additional help. I recommend using the resources on `https://ViewMath.com` You can find video lessons, practice problems, and more. You can also use the online companion for this book to track your progress and access additional resources.

Wishing all students the best in their studies, parents every success in supporting their children, and teachers continued inspiration in their classrooms!

Dr. A. Nazari